Additional praise for *Future State 2025: How Top Technology Executives Disrupt and Drive Success in the Digital Economy*

"The book is extremely relevant as digital has become pervasive throughout business. Every function and process in every industry will be reinvented by technology. If you are not doing it for your company, you'll cease to exist because someone else is doing it at their company."

—Andrew Campbell
CIO at Terex Corp.

"What if you could assemble the brightest practicing CIOs and ask them candid questions about leadership, culture, and strategy? Hunter has done just that. The result is a rare view inside the digital hearts and souls of today's most successful companies."

—Ralph Loura
SVP and CIO at Lumentum

FUTURE
STATE
2025

FUTURE STATE 2025

How Top Technology Executives
Disrupt and Drive Success
in the Digital Economy

HUNTER MULLER

WILEY

This edition first published 2020

© 2020 John Wiley & Sons, Inc.

Registered Office(s)
John Wiley & Sons, Inc., 111 River Street, Hoboken, NJ 07030, USA

Editorial Office
111 River Street, Hoboken, NJ 07030, USA

For details of our global editorial offices, customer services, and more information about Wiley products visit us at www.wiley.com.

Wiley also publishes its books in a variety of electronic formats and by print-on-demand. Some content that appears in standard print versions of this book may not be available in other formats.

Library of Congress Cataloging-in-Publication Data is Available:

ISBN 9781119574798 (Hardback)
ISBN 9781119574835 (ePDF)
ISBN 9781119574811 (epub)

Cover Design: Wiley
Cover Image: © metamorworks/Getty Images

Set in 11/16pt,ITCGaramondStd by SPi Global, Chennai, India.

Printed in the United States of America

SKY10020290_080520

For Brice and Chase

CONTENTS

AUTHOR'S NOTE:

GREAT LEADERS RESPOND AND ADAPT COURAGEOUSLY AND BOLDLY TO DRIVE ACCELERATED TRANSFORMATIONAL CHANGE

Together, all over the world, we are experiencing a historic moment of accelerated transformational change. For many of us in the technology community, the COVID-19 pandemic represents the ultimate challenge of our careers.

The speed of the response is truly extraordinary. In a matter of weeks, we have transformed our organizations and our workforces. These processes of swift change and rapid transformation will likely continue.

What remains unchanged is the essential need for authentic leadership expressing a strong message of hope, confidence, compassion, kindness, and courage.

Over the past 30-plus years, I've been blessed with the opportunity to work with visionary leaders at amazing companies such as Zoom, Darktrace, Adobe, Moveworks, Zendesk, Nutanix, Okta, RingCentral, UiPath, and many others.

Based on my experience, I am absolutely convinced that now is the best time to be a technology leader. Despite the crisis, we are on the cusp of a new era of technology-enabled innovation and reinvention fueled by advances in artificial intelligence, machine learning, 5G, cloud computing, edge networks, virtual reality/augmented reality, facial recognition, and geolocation.

When the scale and severity of the crisis became apparent, our team at HMG Strategy pivoted quickly to shift from a live summit model to a digital summit model. In addition to producing and presenting a full slate of virtual summits featuring top-tier world-class industry thought leaders and executives, we also produced dozens of virtual briefings, meetings, conferences, and webinars.

None of this would have been possible without clear-eyed and confident executive leadership, and a team of dedicated colleagues who shared an optimistic and uplifting vision of the future.

Eric Yuan, Founder and CEO of Zoom, put it nicely when he said, "Hunter and the HMG Strategy team confirm that now is unquestionably the best time to be a technology leader. Executive Leadership Summits highlight the responsibilities and opportunities we have as pioneers guiding the next amazing cycle of innovation on a global scale."

Patty Hatter, SVP Global Customer Services at Palo Alto Networks, told me that "HMG Strategy provides a looking-glass view into the future state of technology and its global impact. Every tech leader needs to be inclusive in thought leadership and HMG Strategy provides the necessary, diverse viewpoints."

I sincerely appreciate the affirmations we have received from our network and I am deeply grateful for their continuing support and engagement throughout these difficult times.

Confronting and Overcoming Unexpected Challenges

I find it fascinating that world-class executives such as Jeff Bezos of Amazon, Bob Iger of Disney, and Mark Zuckerberg of Facebook have resumed their primary leadership roles during the crisis.[1] Modern leadership requires presence, even in a world of virtual meetings. Exemplary leaders actively demonstrate their commitment and sincerity, no matter where they are in the table of organization.

From our unique vantage point at HMG Strategy, we can see the shift and observe how great leaders adjust and adapt quickly to the new normal. Facing a torrent of new and unknown challenges, our leadership capabilities are being tested as never before. From my perspective the new and pressing challenges include:

---

[1]https://finance.yahoo.com/news/why-amazon-facebook-disney-saw-their-chiefs-retake-control-142735464.html

- Leading the digital enterprise courageously and enabling strong growth, inventing new business models and creating opportunities for growth, and profitability in core, adjacent, and new markets.

- Redesigning and reinventing the organization's portfolio of digital assets and capabilities.

- Elevating the customer experience and reducing friction at all levels, in all markets.

- Accelerating cultural shifts through new platforms and collaborative tools that enable, enhance, and augment workforce productivity, anywhere and anytime.

- Ensuring safety and security in all workplace environments and processes, whether at home or in traditional business settings.

- Enabling and implementing processes and technologies required for smooth "return-to-work" scenarios and hybrid scenarios involving a blend of work-from-home and in-office arrangements.

- Identifying and mitigating risk through the next 18 months.

- Inspiring, coaching, mentoring, and guiding the organization through difficult times.

- Creating and supporting an authentic culture of gratitude, humility, collaboration, teamwork, and kindness that will outlast the crisis and serve the organization when new challenges arise.

- Inspiring hope and confidence by providing exemplary and courageous leadership throughout the crisis.

Confronting an onslaught of hard challenges and responding effectively to them will be difficult. Finding and implementing the best processes and solutions will require all of
our skills and experience. They will require the kinds of superior courage, knowledge, and experience that great leaders
bring to the table.

Great leaders accelerate their efforts to reimagine, reinvent, and innovate the customer experience and invent new
business models to drive value across the enterprise while
ensuring the highest possible levels of safety and security.
They deepen and extend the strength of their relationships in
the C-suite and boardrooms, and open new paths to success
in receptive markets. They complete the shift to digital and
embrace the new normal. They thrive and succeed, despite
obstacles and barriers.

At HMG Strategy, we have a history of engagement with
top-tier brands such as Apple, CVS Health, UnitedHealth
Group, McKesson, AT&T, AmerisourceBergen, Ford, Cigna,
Chevron, JPMorgan Chase, General Motors, Walgreens Boots
Alliance, Verizon, Bank of America and Wells Fargo.

I recently asked my friend Snehal Antani to share his
thoughts on leadership excellence. Snehal is co-founder
and CEO of Horizon3.ai, a cybersecurity startup focused on
AI-enabled red teaming. Prior to starting Horizon3, he served
as a CTO for the DoD within the U.S. Department of Defense,
responsible for transforming the technology capabilities of
the command to include cyber security, advanced R&D, AI,
and data analytics.

Here is Snehal's insightful response to my question:

I've worked with both great leaders and terrible "leaders." What makes them different is a question that's been top-of-mind lately. It wasn't about technical expertise, or business savvy, but rather a simple difference: the great leaders were naturally inclusive, making everybody feel that they were part of one team regardless of title, rank, org structure, or pedigree.

With inclusiveness as the foundation, they trusted and empowered their people to move out, giving them room to maneuver, coaching along the way. When things went wrong, or problems were daunting and ambiguous, the truly great leaders were able to roll their sleeves up and "troop lead," quarterbacking their team through issues without seeking blame or fault, but extracting lessons to continuously improve.

They never let issues linger or fester. These leaders inspired me to punch above my weight, helping me discover my limits vs. stifling me. They demanded excellence of me without saying it. They coached and rarely admonished.

The terrible "leaders" I've worked with have been almost the exact opposite: engaging people based on rank, title, pedigree; dictating tasks under the guise of policy or authorities; demanding loyalty; hiding when things are tough; seeking glory off the backs of others.

Great leaders also recognize the extraordinary value and "infinite potential" of their teams, colleagues and partners. "Assume that your team has that infinite potential," says Shankar Arumugavelu, SVP and Global CIO, Verizon. "My leadership style has always been about empowering my teams and giving them challenges. What I want to make sure is that people have the autonomy and I always look for people who are life-long learners."

The Time for Courageous Leadership

Now is truly the time for courageous and visionary leaders to step forward. Here's an excellent example of what I mean by visionary leadership:

I had a wonderful and truly illuminating conversation recently with Nicole Eagan, Co-Founder and Chief Strategy and AI Officer of Darktrace, one of the first companies to develop an AI platform for cybersecurity.

Nicole's extensive career spans 25 years working for Oracle and early- to late-stage growth companies. Named one of HMG's "Top 100 Technology Executives to Watch" in 2019, Nicole has introduced machine learning to enterprises of all sizes. Under her leadership, Darktrace has won more than 100 awards, including being named one of Fast Company's "Most Innovative Companies" and a CNBC "Disruptor 50."

Naturally, we spoke about how the pandemic is transforming the universe of enterprise security and forcing companies to develop new strategies for coping with a continually evolving range of cyber-risks.

"Until recently, the security industry has focused on traditional concepts such as network security, cloud security and endpoint security. Today, with so many people working from home, the focus has shifted to workforce security," Nicole explains.

Here's why: Cybercrime is an organized business operating on a global scale. Attackers are often handed lists of target companies. When most of us were still working from corporate offices, the majority of attacks were thwarted by enterprise-grade cybersecurity systems.

"But all of a sudden, people are working from home on their laptops, using their home Wi-Fi routers, and using a variety of new tools for collaboration and videoconferencing," Nicole says. As a result, the enterprise-grade security systems we had relied on are no longer fully protecting us.

The attackers know this and are probing for new weaknesses they can exploit. "They're circling back to those hardened targets in hopes of finding new vulnerabilities and gaining a foothold," Nicole says. "They're looking for those weak links."

The "new normal" is a constantly changing landscape of cyber-threats and countermeasures. Security strategies must be continually updated, adjusted, and optimized in real time to remain effective.

Cybersecurity solutions that incorporate machine learning can adjust much more rapidly to continually evolving threats and dynamic workforces than systems that are based on "pre-planned" scenarios. "You really can't pre-plan for a lot of this," Nicole explains. "You need the capability to change as the world changes around you."

Another unexpected consequence of working at home is the tendency for work to continue around the clock with no clear downtime. The "always on" nature of work today creates new opportunities for cyber-attackers.

"Not every company is set up for 24/7 operations. That's the reality. So one of the things we've done is extended free 24/7 Proactive Threat Notification coverage to all of our customers for exactly that reason. The most important feedback we've received from our customers is just knowing that in addition to the ML and the AI, our teams of threat analysts are providing 24/7 eyes-on-glass for them. That's really helping them at this time," says Nicole.

I genuinely appreciate how Nicole and her colleagues at Darktrace have developed such a clear vision of the present moment and its unique challenges. From my perspective, Nicole is a role model for technology leaders and executives everywhere.

Pivot Quickly and Effectively

As mentioned earlier, our team at HMG Strategy pivoted rapidly to a digital summit model. Without missing a beat, we produced a full slate of virtual summits, drawing on the strength of our global network to feature an all-star roster of thought leaders and executives.

Here's a brief sample of the genuinely valuable insights offered by some of the exemplary speakers, presenters, and panelists at our virtual summits:

Wendy M. Pfeiffer, CIO, Nutanix: "The biggest lesson for me over the last two weeks is how to help people create structure when they are working completely remotely. The kneejerk reaction is to feel like it's an emergency and to work 24/7. But that will exhaust people. So, what I've done with my team and we've since rolled it out to many employees in the company is that I've asked people to block out their work hours so that they can be available to help their kids with homework or if they need to take a walk.

What we discovered is that when we didn't force people to fit into traditional work hours, we found more capacity for people to be available for extended hours in exchange for taking some hours to be available for their families. I'm encouraging people to do this. Even creating that schedule week by week is extremely helpful from a psychological standpoint.

I see each member of our C-suite not only connecting with each other but also reaching out to schools and their communities to help them to connect during this crisis."

Steve Phillpott, CIO, Western Digital: "This over-rotation to working remotely will forever change our work environments and we're not going to return to work as it was before.

As we manage through times of crisis, it accentuates the importance of communication and the robustness of our communication and collaboration tools. Now more than ever, communication is critical, and we need to communicate more than we think we need to, and we need to seek out new forms of collaboration. As we're working remotely, we need a robust set of tools that can scale. On our side, we've seen a 3–5X increase in collaboration tool capacity in days, so the ability to scale quickly is critical ... IT must serve as a role model to the rest of the company as a remote workforce."

Bhavin Shah, CEO & Founder, Moveworks: "Each of us are working remotely and we're seeing increased demand for IT services, including a 2X increase in overall IT support ticket volumes.

Every walk-up help desk in America is vacant—they can't be used. So, we're seeing a 3X increase in demand coming through the chatbot using these digital channels. Systems and access have gone through the roof ... we're seeing greater demand for automation, to use AI and machine learning to resolve support issues. And this shift to remote workforces has become a semi-permanent thing that America and the world is experiencing.

And while many of us will eventually return to the office, we'll gain a deeper appreciation of what it takes to provide that type of support remotely and digitally.

It's important to give people an understanding of how the landscape is shifting. We're helping organizations to quickly transform and become better at supporting employees as they are remote. We're seeing a massive uptick in chat adoption. How you operate in this mode is going to separate those companies that succeed going forward and those that don't."

Do You Inspire Confidence and Trust? Are You a Beacon of Hope and Compassion?

Everyone has their own ways of coping with stress. When you're an executive or leader, however, you have a special responsibility to help others manage through a crisis. You have to be more than a business leader; you need to be a role model for the behaviors that will most likely result in positive outcomes for the organization and the people in it. You must become a beacon of hope and compassion.

In difficult times, great leaders inspire confidence and trust. They provide guidance, counsel, and advice. They listen and they learn. They become true servant-leaders, serving the larger needs of their colleagues and co-workers.

Is the glass half empty or half full? I bet that most of us in our amazing industry would say it's half full. From the perspective of the modern technology leader, we have been given the rare opportunity to genuinely transform ourselves and our companies.

Let's not waste this opportunity. Let's learn from it, let's build on it, and let's collectively find the good in these difficult times. I firmly believe that by working together and leveraging our incredible talents and abilities, we will emerge stronger and better from this crisis.

Today, modern technology leaders everywhere are playing pivotal roles in determining the future of the enterprise. I am exceptionally confident that as a group, we will provide the guidance, counsel, advice, and experience required to drive forward and achieve success.

Remember, the enterprise is counting on us to overcome the hurdles and devise practical solutions on both the tactical and strategic levels. This is the true test of our character and our determination. I know in my heart that we will make the proper choices and navigate successfully through this storm.

Are You Being Your Best Self?

In one of our early virtual events, my good friend Tony Leng observed that great leaders are "givers, not takers." He reminded us to remain authentic and to offer hope, reassurance, and confidence to all of the people who look to us for guidance and direction.

In my conversation with Tony, we spoke mostly about the need for honest and courageous leadership in times of extreme emergency. Here's a quick summary of the advice we shared for technology leaders and executives:

- Be authentic when you're leading a virtual meeting. People can sense authenticity, and they're looking to you for reassurance. Great leaders radiate authenticity and honesty, even in virtual meetings.

- Be a giver, not a taker. Again, people can sense your intentions. Show them you have a big heart and that you are prepared to make sacrifices for them.

- Hold virtual meetings frequently and keep them short. Assign relatively simple tasks and quick projects to keep your people engaged in their work. This is a time for using the light touch. It's okay if the meetings are brief; what's important is regular check-ins to touch base and reassure people that their contributions still matter and that you still value their work.

- Remember that the crisis will end and that when it's finally over, the quality of your leadership will be judged by how you acted in these difficult times. Make no mistake: The coronavirus will likely be the ultimate test of your leadership capabilities. Be your best self and don't be afraid to show your compassion.

I wholeheartedly agree. As business and technology leaders, we definitely have the strengths and experiences required for handling periods of turbulence and uncertainty. As a group, we have a unique view of complex transformational processes. We know how to get things done and we're not afraid to tackle hard problems. Together, we will prevail and emerge stronger.

Hunter Muller

May 15, 2020

ACKNOWLEDGMENTS

This is my sixth Wiley book and, like its predecessors, this book represents several years of in-depth research and analysis.

I am honored to acknowledge the contributions of our expert sources who shared their time, insight, and experiences freely and without reservation. I am deeply grateful to Eric Anderson, Snehal Antani, Renee Arrington, Shankar Arumugavelu, Ashwin Ballal, Shawn Banerji, JP Batra, David Bessen, Steven Booth, Chris Colla, Bob Concannon, Barbara Cooper, Jamie Cutler, Dale Danilewitz, Daniel Dines, Amy Doherty, Nicole Eagan, Mark Egan, Thomas Farrington, John Fidler, Hugo Fueglein, Curtis Generous, Chuck Gray, Marc Hamer, Jon Harding, Vishwa Hassan, Patty Hatter, Jason Hengels, Donagh Herlihy, Sheila Jordan, Chad Kalmes, Tom Keiser, Steven Kendrick, Justin Lahullier, Zackarie Lemelle, Tony Leng, Beverly Lieberman, Ralph Loura, Christopher Mandelaris, Quintin McGrath, Lynn McMahon, Matt Mehlbrech, Vipul Nagrath, Vish M. Narendra, Earl Newsome, Helmut Oehring, Tom Peck, Wendy Pfeiffer, Steve Phillips, Steve Phillpott, Mark Polansky, Phyllis Post, Nathalie Rachline, Dan Roberts, Dave Roberts, Jon Roller, John Rossman, Dr. Kenneth Russell, Vijay Sankaran, Candida Seasock, Bhavin Shah, Jikin Shah, Naresh Shanker, Prabhash

Shrestha, Stephen Spagnuolo, Patrick Steele, Milos Topic, Angie Tuglus, Sangy Vatsa, Ken Venner, Gautam Vyas, Craig Walker, Bart Waress, Patricia Watters, Jon Wrennall, Angela Yochem, Eric Yuan, and Jedidiah Yueh for contributing their expertise and wisdom to this book.

I also thank my excellent team here at HMG Strategy: Kimberly Ball, Conor Claflin, Melissa DiPreta, Travis Drew, Jef Fisk, Nikki Frederick, Tom Hoffman, Rob Kovalesky, Melissa Marr, Bryan McCreedy, August Pelliccio, Mark Pelliccio, Peggy Pedwano, and Lindsay Prior.

Additionally, I thank the members of our editorial group: Tom Hoffman, Mike Barlow, and August Pelliccio. Their contributions were truly indispensable.

And of course, I thank my editor, Sheck Cho, for his continued support and guidance over the years. I am proud to be a Wiley author and sincerely appreciative of the role played by Wiley in our industry.

PREFACE

The term "perfect storm" typically refers to a rare combination of catastrophic events leading to an unimaginably awful outcome.

In the technology industry, however, we are experiencing the opposite. Now is the perfect moment to be a technology executive. Collectively, our timing could not be better.

I am truly fortunate to have a front-row seat for this remarkable convergence of amazing leaps forward in the rapid development of exciting new technologies, systems, and platforms.

We are living at an unusual point in history. From my perspective, our industry has achieved phenomenal success. Many of our dreams have been realized and are now woven into the fabric of our daily lives.

Moreover, we are on the cusp of even greater achievements, thanks largely to a continual stream of innovation in numerous fields, such as artificial intelligence, machine learning, natural language processing, robotic process automation, cloud computing, high-speed data networks, 5G wireless communications, connected devices, low-cost sensors, distributed ledger, virtual reality, augmented reality, advanced manufacturing, transportation, and energy.

Those continuous waves of innovation can exert a heavy toll on the modern enterprise. Sometimes new technologies can be tempting, even if they are not a good fit for your business model. And even if they are great for the business, they might not integrate smoothly with existing infrastructure.

"You don't want to throw a bunch of digital assets against a wall and see what sticks," says Ralph Loura, senior vice president and chief information officer at Lumentum, a market-leading manufacturer of innovative optical and photonic products enabling optical networking and commercial laser customers globally. "You need to be disciplined and stay focused on execution. Technology should create value, for the customers and for the business. If the technology isn't connected to your value stream, it's not likely to generate a return on your investment."

I like how Ralph cuts directly to the chase. It is all about creating value for the enterprise and its markets. Plain and simple.

At the same time, I also believe that technology executives have an obligation to push the envelope and help the enterprise go where it has never gone before.

Executives should have a strong focus on raising a company's earnings before interest, tax, depreciation, and amortization (EBITDA), according to Marc Hamer, a chief information officer and chief digital officer currently working in private equity. Marc was formerly a corporate vice president, chief information officer, and chief digital officer at Sealed Air, an innovative packaging company known for its Cryovac food packaging and Bubble Wrap cushioning packaging brands.

"You have to think of yourself as an accelerator to everything the business wants to do. Use technology to get back time, to go faster, to solve more," Marc says. "If you can do that, your organization will be the most sought after in the industry and provide enormous returns."

Unquestionably, the rise of cloud computing has made the technology executive's life more interesting—and more challenging. Everyone is heading for the cloud. But not everyone knows what to do once they have gotten there.

Wendy Pfeiffer is an authentic cloud expert. She is the chief information officer at Nutanix, an amazing company that unifies private, public, and distributed clouds and empowers IT to deliver applications and data that power their businesses. Nutanix is a leader in hyperconverged infrastructure appliances and software-defined storage.

"The newer technologies empower you to make small changes that have large effects," Wendy says. "We've moved past the era in which you had to spend millions of dollars, or sometimes tens of millions of dollars, to make a significant change. The new tech allows you to operate at an atomic level. You don't have to build a new CRM system or invent a new taxonomy for your information. Today's tech is much more subtle, and tremendously powerful."

Wendy really hits the nail on the head. Yesterday's tech was like a giant freight train while today's tech is more like a Formula One racecar. For technology executives, this translates into vastly higher levels of agility and incredibly shorter business cycles. Essentially, you have got your foot on the accelerator throughout the race.

Daniel Dines is the founder and chief executive officer of UiPath, a global industry leader in robotic process automation (RPA). I spoke with him just before the manuscript for this book went into production, and I am delighted that we were able to include his valuable insight. In our conversation, I asked Daniel to describe how RPA is transforming the modern enterprise. Here is a brief summary of his reply:

> Most agree on the importance of AI. But operationalizing AI is a challenge. Your data scientists can write beautiful predictive models based on machine learning, but getting those models into production isn't easy. By automating the mundane and repetitive tasks, RPA frees up time for your data scientists and lets them do what they're good at doing. So we see RPA enabling AI in the enterprise. It's an absolutely critical step on a long journey that is unfolding very fast.

From my point of view, Daniel is a brilliant leader and pioneer. Innovative thinkers such as Daniel are rocking our world—in a good way.

A Shift in Focus

In this wild and turbulent world, what keeps the modern technology executive up at night? It is not the email servers or the ethernet connections. Those days are long gone.

Today's technology executives focus on providing the leadership and guidance required to lead, innovate, disrupt, reimagine, and reinvent to create and sustain competitive advantages amid rapidly evolving markets and constantly changing customer expectations. That is the challenge—now and for the foreseeable future.

Recently I wrote about the problem of legacy tech companies struggling to keep pace with rapidly evolving markets, shifting customer expectations, and an absolutely astonishing array of products and services based on newer and more powerful technologies.

From my perspective, we have moved way past the tipping point. This is a truly watershed moment for the technology industry and the larger global culture. For years people have been talking and writing about the "new industrial revolution," and now it is happening, in real time.

For many of the legacy tech firms, however, the big question is whether they are active participants in the revolution or merely spectators watching from the sidelines. That is not an idle or trivial question; their futures depend on their ability to keep up with global markets that are moving ahead and continually changing with incredible speed.

But instead of focusing on legacy firms, I prefer to spend my time with the bold and innovative leaders of companies that are proactively shaping the future of our industry.

At HMG Strategy, we are fortunate to have sponsor partners that are genuinely blazing new trails and pioneering new business models across the expanding technology universe. Our partners include these superlative forward-thinking companies:

- Nutanix, whose Enterprise Cloud OS software delivers one-click simplicity to infrastructure and application management, elevating IT focus

- Zoom, a leader in modern enterprise video communications, with an easy, reliable cloud platform for video, audio, and more

- Lenovo, a Fortune Global 500 company focused on the bold vision to deliver smarter, world-changing technology for all

- UiPath, which is leading the "automation-first" era by delivering free and open training and through its robotic process automation (RPA) platform

- RingCentral, a global provider of flexible, cost-effective cloud enterprise unified communications and collaboration solutions

- Adobe, whose tools help customers create highly compelling content and whose innovations drive the future of digital media

- Google Chrome Enterprise, which delivers a future-proof OS that makes it simple to work smart and keep your business safe

- Darktrace, a leading cyber-AI company whose "Autonomous Response" technology is modeled on the human immune system

- Fortinet, which provides top-rated network and content security as well as unique security fabric–based secure access products

- Moveworks, which provides an AI platform purpose-built to solve one issue for large enterprises—resolving their employees' IT support issues

- Info-Tech Research Group, the fastest-growing and most innovative IT research and advisory group, serving 30,000 IT professionals

- Zerto, whose software platform simplifies workload mobility to protect, recover, and move applications freely across hybrid and multi-clouds

- Digitate, a software company that leverages machine learning and AI to manage IT and business operations

- Equinix, a global interconnection platform linking the world's leading businesses to their customers, employees. and partners

- Zendesk, a powerful and flexible customer service and engagement platform that scales to meet the needs of any business

- Ivanti, which unifies IT and security operations to better manage the digital workplace, reducing risks with insights and automation

- Pure Storage, a data platform powered by all-flash storage, offering a simple, effective way to build a better world with data

- Box, a cloud content management company that empowers enterprises to revolutionize how they work

- Delphix, whose Dynamic Data Platform provides an approach to DataOps that enables easy and secure data delivery

- OutSystems, the number-one low-code platform for rapid and simplified application development, with high attention to detail

- Catalyte, an AI company that solves a superlative challenge in IT today: creating a diverse, affordable, and sustainable ready-to-hire workforce

These companies are role models for a new generation of tech firms. I am deeply grateful for their presence and involvement in our summits, conferences, and events. From my point of view, they represent the future of our industry. They are lighting up a path for all of us and clearly demonstrating how continuous innovation and disruption are essential to business growth and success.

I sincerely hope that you enjoy this book and that you take advantage of the advice gathered from dozens of experts and industry thought leaders. I hope to see you at one of our summits, and I look forward to hearing your ideas and suggestions for achieving the future state of technology for the modern enterprise.

FUTURE
STATE
2025

INTRODUCTION:

REIMAGINING AND REINVENTING THE DIGITAL ROADMAP

Today's leaders need a new playbook and a reimaged digital roadmap to prepare for the future state of technology and business.

Technology leaders must be ready to lead, innovate, disrupt, and invent solutions to grow the business in core, parallel, orthogonal, and new markets. They also must be ready to prepare themselves for roles in the corporate boardroom.

This book is the roadmap to success in the 21st century. It is the essential and indispensible guide for technology leaders and executives in the modern global enterprise.

Facing the Challenge

The modern enterprise needs a bulletproof digital strategy to guide its journey through unprecedented times of radical change and transformation. The customer experience has become the most powerful driver of growth and revenue, forcing organizations to rethink and reinvent their go-to-market strategies.

Today's digital enterprise must deliver continuous innovation and invention—or risk becoming irrelevant. It is absolutely imperative for senior executives and boards to select and deploy the right technologies at the right time. Staying ahead of the technology curve leads to growth and success; falling behind leads to obsolescence and failure.

Winning Strategies for Success

This book is written to help executives identify the best technology investments and move forward with rapid implementations of new systems and solutions—ahead of the competition!

History is merciless. Yesterday's winners are today's losers. Since the inception of the Dow Jones Industrial Average in 1928, its list of companies has changed 51 times. Which side of the fence are you on? Have you done everything you can possibly do to prepare your organization for success? This book will help you make the right choices.

How This Book Is Different

This book is a roadmap to success through technology innovation and invention. Using our unique in-depth interview process, we focus directly on the practices and habits of leading technology executives at the world's largest and most successful organizations.

This book includes in-depth first-person interviews with leading technology executives, explaining and sharing their strategies for acquiring or developing new technologies, systems, and platforms. It offers a unique, world-class playbook for continuous innovation and invention.

Specifically, readers will learn how the world's leading executives select, develop, and implement the newest techniques and technologies, including:

- Artificial intelligence (including machine learning, deep learning, reinforcement learning, neural nets, natural language processing, and cognitive computing)

- Automation and robotics (including design, integration, control, and optimization)

- Competing for talent (including recruitment, hiring, training, retention, and incentives)

- Advanced cybersecurity (including continuous monitoring, war games, and proactive threat hunting)

- Extended reality (including virtual reality, augmented reality, and 3D gaming)

- Customer/User experience (including design thinking, rapid prototyping, development, testing, deployment, innovation, and improvement)

- Cloud strategy (including public, private, and hybrid)

- Autonomous transportation and logistics (including cars, trucks, vans, shipping, heavy rail, light rail, and drones)

- Resiliency (including predictive maintenance, disaster recovery, and operational readiness)

- Smart cities (including the Internet of Things, ambient computing, continuous surveillance, facial recognition, voice analysis, emotional state recognition, and privacy)

Unrelenting Pressure to Innovate

Companies face excruciating pressure to create and deliver innovative business strategies for growth and success in rapidly changing competitive markets.

Expectations have risen dramatically: In today's digital economy, the customer experience must be fast, flawless, and completely secure.

Are you up for the challenge? Do you have the technology and talent in place to meet the expectations of the C-suite and board of directors?

Speed, Agility, and Creativity

For the 21st-century enterprise, success will depend on speed, agility, creativity, and excellence at all levels. But achieving phenomenal success will require more than just strong business acumen and superior leadership skills; it will require true digital fluency and an extraordinarily deep understanding of new technology.

Future State 2025 provides a meticulously detailed roadmap to success in the age of digital transformation. It is written expressly for top executives, board members, investors, innovators, and entrepreneurs. It is a uniquely valuable collection of insight, experience, and firsthand knowledge, collected from the very best minds of our generation and gathered into one absolutely essential book.

I invite you to join me on a journey of innovation, invention, and discovery. Your wisdom, experience, and insight are essential to the global collaborative process that enriches the lives of people all over the world.

Critical Insights

Future State 2025 dives deeply into the critical imperatives defining the CIO's mission in the modern enterprise:

- Reimagining and reinventing the digital customer experience to deliver value at every touchpoint across the enterprise
- Enabling innovation and empowering the C-suite's vision for growth in core, parallel, adjacent, and new markets
- Security across the digital enterprise
- Hiring and inspiring top talent
- Simplifying technologies and making them easier to use
- Building strong partnerships with strategic vendors

- Creating an agile culture focused on speed-to-cash, performance, and innovation

Future State 2025 is the essential roadmap for successful leaders worldwide, a landmark book for top executives, boards, and investors in all industries and economic sectors.

Future State 2025 is my sixth book on the critical subject of technology executive leadership. It introduces a game-changing and truly visionary approach to the challenges of integrating innovative technology and modern business strategy to generate value for the 21st-century enterprise.

Chapter 1

Technology Leadership Is Indispensable and Essential

Now is undoubtedly the best time to be a transformational technology leader. We face challenges, but we also have unprecedented opportunities for leading and facilitating innovation, disruption, and growth in core, parallel, and new markets.

That's why it's absolutely essential for technology executives to choose the right technologies and make the wisest investments. The future of their organizations depends largely on the choices they make today.

I predict the technology industry will continue to prosper and grow. Today everyone is a technology consumer, and that trend shows no signs of slowing down. In fact, every reliable indication points upward. IDC anticipates $7 trillion in IT-related spending in the 2019 to 2022 time frame.[1]

In their anxiety over future earnings, investors often make poor choices. Sometimes fear of the unknown overcomes rational instincts. As technology leaders, we simply cannot afford to make decisions based on momentary events and temporary setbacks. We must think in *strategic* terms. That is both our role and our responsibility.

The economy itself is strong, and tech remains a driving force for growth in markets all over the world. Though there are signs on the horizon of an economic slowdown, a major recession appears unlikely.

Long-range forecasts are notoriously inaccurate, and it seems like a waste of energy to fret about events that might or might not happen 12 months down the road.

Here are some predictions for the shorter term:

- The need for cloud storage will accelerate and continue growing.
- Cybersecurity will remain a major challenge for companies of all sizes.
- The war for talent will make it harder to hire and retain the best employees.

- The shift from traditional IT to infrastructure as a service will continue as more companies seek to reduce capital expenditures.

- Achieving excellence in IT executive leadership will remain a top priority at forward-thinking organizations.

For those of us operating within the tech industry, it's hard to be a pessimist. From our perspective, the future seems bright and full of promise. That doesn't mean there won't be disappointments and bumps in the road. That said, it is no longer possible to imagine a world without digital technology.

Technology has become an economic necessity at every level. You simply cannot run an organization of any kind without technology. Soon the majority of the world's citizens will be digital natives. For them, technology is a basic right, like breathable air and clean drinking water.

I am a realist, not a starry-eyed optimist. I firmly believe we are in the opening innings of a global transformation. We have a long way ahead of us, and technology will continue playing the dominant role in defining our future as a global society.

Role of the CIO in Guiding the Enterprise to the Future State

One of the greatest challenges that corporate executives face these days is positioning their companies to survive and thrive in the next three to five years and beyond. According to the *Harvard Business Review*, 52 percent of the companies that were in the Fortune 500 in the year 2000 have gone bankrupt, been acquired, or ceased to exist due to digital disruption.

Because of their unique view across the enterprise and outside of the organization, CIOs are well positioned to help the CEO, line-of-business leaders, and fellow members of the C-suite to identify and act on new business opportunities that can deliver new customer experiences and help differentiate the brand.

A good starting point for CIOs and technology executives involves fostering innovation among their teams and with key stakeholders in the enterprise. "It's critical for CIOs to encourage innovation and out-of-the-box thinking," says Vishwa Hassan, director of Data and Analytics at USAA.

Because innovation often stems from failures, it's also important to reassure team members that failure can be accepted—so long as the organization learns from its flops. "It's important to reward failure—not for a big project failure but for failures that occur early on in the life cycle so that learnings from those failures can be incorporated into key attributes of what doesn't work," Vishwa adds.

Vishwa also recommends offering process- and standards-based solutions for ensuring enterprise stability while being nimble enough to demonstrate quick turnarounds for revenue-generating projects in sales, marketing, logistics, and other areas in the company.

Meanwhile, when it comes to assessing advanced technologies that can help provide the organization with a competitive edge, Vishwa points to the use of operational analytics as a potential differentiator. "I'd recommend focusing on analytics that in near-real-time impact operations compared to a dashboard or model output that a group has to review and implement actions based on those results."

Tailoring the Message for Board and C-Suite

As CIOs and technology executives spend an increasing amount of time presenting to their boards of directors, they're discovering how the messaging needs to be tailored to meet the needs and interests of board members. Indeed, 78 percent of CIOs say they are communicating with the board more than ever before, up from 67 percent in 2018, according to IDG's 2019 State of the CIO survey.[2]

Most board members are interested in the business impact of technology investments, the costs associated with those investments, and the inherent risks associated with implementing and applying technologies. Since most board members aren't technologically savvy, they're not interested in highly technical discussions.

I caught up recently with Dale Danilewitz, EVP and CIO at AmerisourceBergen, to capture his approach to connecting effectively with the board and for moving the business forward. Here's a lightly edited transcript of our conversation.

Hunter Muller: How do you lead into the C-suite and help the board understand how to digitally connect with the customer to reimagine and reinvent the customer experience?

Dale Danilewitz: I think the short answer is that my CEO is productively paranoid. He is constantly concerned about what's next on the horizon and the various forces that may impact us or create obstacles for us to deliver on our goals. For me, it is therefore not about a decision or strategy to open ourselves up to the prospect of disruption. It should be inherent in one's culture. I'm a big believer that technology is not the catalyst to raising the prospect of disruption, it should be one of the many industry forces that can influence a company's path to deliver on its strategy, both as an enabler and as a competitive inhibitor.

One should regularly be using scenario planning while engaged developing and revising one's strategic direction. This includes both evolutionary and revolutionary planning. If there's a trigger point that indicates the strategy is not delivering, you must be courageous to change or pivot. We now call this disruption, but I believe that disruption has been occurring since the inception of business, and technology is more of an accelerator, introducing disruptive forces more rapidly than in the past.

One can't allow oneself to plan within the confined parameters that one applied in the past. One needs to elevate oneself to ensure one understands the industry and societal forces that are shaping the business landscape. Peter Schwartz introduces the concept of scenario planning in a 1996 book called *The Art of the Long View*, where he uses the rise of the oil giant Dutch Shell Oil to illustrate the benefits of this approach.

HM: The Fortune 500 will turn over even faster now because of technological disruption.

DD: Technology is a contributor to shifting your industry. If you have a culture of strategic planning where you are constantly accounting for all variables that could influence your future, then you're in a much better position to factor in the advances of technology and the opportunities that they offer.

Yesterday I was talking to someone about board positions for IT professionals. It's interesting because when you look around, the number-one priority considered by boards when it comes to information technology is cybersecurity, and that's always something I'll address with my board. Many boards and board members are nontechnical, which requires us to explain technology in the context of business opportunity and risk. They are also very well read and expect us to discuss familiar technical terms and their application in our business.

We have chosen to host a tech fair at our next board meeting where we will bring in many of our innovative applications of technology that we are either applying commercially or piloting for future possible use. Our goal is to give them an appreciation of the investments we are making in exploring emerging capabilities delivered by the latest technologies. Board members want to know from a governance perspective that we are looking ahead while continuing to operate our business.

HM: How do you go to achieve that broad reach? Partnering with VCs and tech startups in Tel Aviv or Silicon Valley?

DD: It's all about business enablement and driving our strategy to meet our vision. I intentionally stay away from talking about technology in our board meetings. The board is aware of what we're doing—we are using the tech fair to inform them of what we are doing to ensure we aren't lagging when it comes to using innovative and emerging technologies (e.g., piloting blockchain projects, augmented reality, AI, and drones) while sharing the work we've done in design thinking and how we have leveraged these concepts to get closer to our customers. As for obtaining exposure to innovation and opportunity, we partner with accelerators and innovation centers to ensure we have a front seat to early entrants in developing technologies and capabilities.

HM: How do you go about inspiring your team to think differently?

DD: My advice to our colleagues is when we talk about capabilities, we try to focus on the problems we are trying to solve and not fall into the trap of a "solution looking for a problem." We also ensure we are considering both the business unit and the enterprise across our business units to leverage solutions while not suppressing local innovation. My job is to look across the enterprise to identify these opportunities and allow my business partners to focus on their specific areas. We run hack-a-thons both within our business units and across the enterprise and ensure we have cross-pollination of ideas as well as brainstorming vertically and horizontally. We have also accelerated our migration into fully agile product teams across the enterprise.

HM: How do you view the current market for technology talent?

DD: There's not as much of an incentive for foreign students to stay behind to work in the United States after they graduate. We need to find a way to encourage them to be added to the local talent pool by helping them ease into permanent residence. Home countries like India and China are making it more attractive for them to return and thus competing with us for the same talent. This is contributing to the dearth of available talent in the United States.

I remind my team, "When interviewing a potential candidate and you feel strongly about them, be ready to provide them with an offer letter before they walk out the door so that they feel wanted and we don't lose them to the market."

My concluding thoughts relate to a panel question at a national conference where we were asked about the pressure and anxiety we face during these times of "shadow IT" being more pervasive and tech startups waiting to eat our lunch. Some of the panelists bemoaned our current state of affairs; however, the most profound and appropriate comment came from the CIO sitting next to me when he raised his eyebrows and blurted out that it was the greatest time to be in our position as corporations are turning to us for opportunities to drive competitive advantage. I agree; we should feel pretty good at a time when digital solutions are dominating the airwaves.

Pressure Testing Candidates for Executive Leadership Roles

How do you define the characteristics of great leadership in the modern corporate organization? That's a question that invariably arises during my discussions with top executives at major competitive firms and companies all over the world.

Today a large part of leadership is strategic. The best leaders do more than deal with current-state problems. They see around corners, look over the horizon, and develop a clear vision of the future. Then they act boldly and courageously to bring that vision into reality.

At a December 2018 HMG Strategy Financial Services CIO Summit in New York City, I had the honor of moderating a truly brilliant panel of experts in the field of executive search. The panelists included Renee Arrington, president and COO, Pearson Partners International, Inc.; Chuck Gray, consultant, Egon Zehnder; and Stephen Spagnuolo, managing director, Digital Security & Risk, Quantum Search Partners.

I asked them to describe the qualities of great executive leaders. Here's a very brief topline summary of useful insights they shared with our audience:

- Great leaders focus on identifying and solving strategic challenges.

- Great leaders always look for new ways to create revenue and drive business growth.

- Great leaders pay attention to current events and know what's happening outside their organization.

- Great leaders leave behind a legacy of great teams.

- Great leaders understand how their companies make and spend money; they know how money flows in and out of the organization.

- Great leaders have genuine intellectual curiosity.

- Great leaders understand the value of reverse mentoring as a technique for staying in touch with latest trends and shifting markets.

- Great leaders understand why innovation is absolutely essential in modern ultra-competitive markets; they learn how to "fail fast, fail cheap" when developing new products and services.

Additionally, great leaders understand that technology has become a strategic weapon. That concept of "technology as a weapon" is radically different from the traditional notion of IT as a back-office function. Today technology is front and center—you simply cannot compete without superior technological capabilities.

World-class leaders convey their ideas and describe their goals in terms that are crystal clear and easily understood by everyone in the enterprise. In other words, they are totally honest and forthright. They work hard to avoid confusion and misunderstandings. Great leaders strive for clarity, effective collaboration, and a shared sense of purpose.

Building Your Personal Brand

Now is truly the optimal time to be a technology leader. Analysts and experts agree that technology is driving strong economic growth and fueling prosperity. This is our moment, and we are truly fortunate.

That said, none of us can afford to rest on our laurels. We all need to build our personal brands and enhance our reputations as leaders. The process of brand building is continual. It requires our attention and our energy.

In advance of the 2019 Silicon Valley CIO Executive Leadership Summit in Menlo Park, California, I spoke with four experienced technology executives and longtime members of the HMG Strategy global community. Here's a selection of their expert advice for building and nurturing your personal brand:

"Think of a brand you love, such as Pepsi or Nike," says Ralph Loura, SVP and CIO at Lumentum. "What do those brands stand for? Think about yourself and talk to business unit peers or former colleagues and ask them what they would say about you. If they say they see you as transformational, that's great. If you're described as the guy who keeps the projector going, take steps to change that."

Ralph is one of our network's most articulate and thoughtful executives. In a recent conversation with our research team, he agreed that building your personal brand is more important than ever before.

"From a role perspective, we all get typecast. You're either a transformational leader or good at containing costs or deploying ERP systems. So, if you want to change your brand, you need to change how you're being perceived," he says.

Mark Egan, partner at StrataFusion, recommends focusing on your core strengths and using them to enhance your reputation as a thought leader. "I encourage people find relevant topics and then to write blogs and speak at events as much as possible," he says. "There are so many opportunities for professionals like us to demystify technology and simplify complicated topics like cybersecurity."

Successful leaders work on their personal brands habitually. "Don't wait until you're looking for a job," says Mark. "You should be building your personal brand and extending your network continually. The HMG Strategy events are great opportunities for brand building and learning."

Brand building involves getting out of your comfort zone and listening to what other people are saying. Seasoned thought leaders spend most of their time listening and learning. "Find out what people are interested in and learn more about their points of view," Mark advises.

Social media creates an atmosphere of transparency that some find uncomfortable. Smart leaders, however, leverage social media to expand their influence and build up their credibility.

"The number-one rule is to do your job first. The more you excel at what you do, the more your brand will shine. For internal branding I recommend getting out of your comfort zone and getting involved in your business with your business peers. Being seen as someone who goes above and beyond the duty of their day-to-day job and gets actively involved with the business always bodes well internally," says Patrick Steele, chair of the CIO Advisory Board at Blumberg Capital.

From an external perspective, Pat recommends getting active in community service and serving on boards. "The exposure to others and outside issues helps you grow and broadens your experience. I also recommend being active in industry association events. If there is an HMG Summit taking place near where you are, getting involved in planning and presenting is excellent for your career growth."

Jon Roller, CIO at Horsley Bridge Partners, also highlights the value of industry events and conferences. "Industry conferences are great places to meet people who may be on similar career paths. Treat those conferences not only as learning experiences but as branding exercises. Get to know panelists and volunteer to be on panels. Find professional organizations that you are interested in. Those organizations allow you to get to know like professionals and build real relationships. Your brand and network is going to help drive your opportunities," says Jon.

In addition to attending conferences and summits, it's important to expand your knowledge of the business. "Understanding the business side of the company is absolutely essential. You can't just be a technology person. You have to understand sales and marketing and how they drive the business. Same with finance and product," Jon explains.

In many respects, brand building is a discipline requiring constant practice. Jon says, "Always work on yourself. Don't expect the world overnight. It takes time and work ethic to build your brand. But it's an essential element to career ascent."

Learning Valuable Lessons from Customer Centricity

One of the best things about the current technology era is how companies and their teams are creating user-centric experiences that make it easier for executives, managers, and employees to do their jobs. Technology vendors are embracing the "consumerization of IT" by which they are striving to more deeply understand the needs and preferences of their clients and design more user-friendly interfaces and experiences when compared to legacy applications and systems.

The ramifications for CIOs to deliver on these rising user expectations are significant. In a 2016–17 Deloitte survey of executives on the topic of IT leadership transitions, 74 percent of respondents said that CIO transitions typically occur when there is a general dissatisfaction among business stakeholders with the support that CIOs and their IT teams provide.[3]

Chief among these are communications technologies that enable today's highly distributed workforce to communicate and collaborate more closely with one another. Unquestionably, advanced communication technologies are critical for connecting dispersed project teams and for strengthening employee engagement and performance; research has shown that up to 90 percent of communications is nonverbal. Even in today's digital workplace, face-to-face communications have become more important than ever.

One technology vendor that's squarely addressing the demand for face-to-face communications is Zoom, the leader in modern enterprise video communications.

I spoke recently with Eric S. Yuan, Zoom founder and CEO, and asked him about the importance of communicating the vision for delivering state-of-the-art employee collaboration and video experiences to the enterprise. Here is a lightly edited transcript of our conversation:

Hunter Muller: How is Zoom striving to reimagine the employee collaboration and video experience?

Eric Yuan: Look at the trends in the market. More and more teams are virtual (63 percent of companies today have remote workers, according to Upwork).

Second, traditional offices are going away. Third, more than one-third of workers in the United States are Millennials. Collaboration is becoming more and more important in project-focused organizations. We need to empower those employees to collaborate and to get things right.

The work we're putting into the Zoom platform is key to addressing these trends. Plus, we realize now that we not only leverage Zoom to collaborate but that employees are more engaged during video discussions instead of multitasking during conference calls. On Zoom, you're able to see each other and make that visual connection.

HM: How do you communicate your vision for achieving the art of the possible with employees at Zoom?

EY: We have ten offices worldwide. At most, I only travel once or twice a year but I'm using Zoom every day to communicate with employees around the world.

By regularly using Zoom, I'm able to demonstrate to employees what's possible with the platform. If I'm in

California and you're in New York, we're able to shake hands using the platform. It's reliable and easy to use and employees have counted on that.

For companies to succeed, trust is everything. It's absolutely essential to provide employees with a trusting environment. Trust is the foundation for speed and for moving forward.

It's very important because your employees may not all be in the same offices—they could be in remote offices, there are home workers, etc. Email is not an intimate experience, and by human nature we often multitask. Video communication is very different—it's more engaging and a more personal form of communication and it helps to strengthen productivity.

HM: How would you characterize your leadership style in galvanizing the team around your vision?

EY: I feel my style is pretty straightforward. I truly care about our employees, our team, our company, and our customers. "What more can I do?" is a question I constantly ask myself. Open communications is a given in business today.

HM: What are the top challenges Zoom has faced in its go-to-market strategy over the past few years?

EY: The number-one challenge is around the brand. Even if you have a great product, if a prospective client has never heard of Zoom, why would they take the risk of adopting our platform? The only time we've ever lost a deal is when a prospect tells us they have no interest in changing.

The number-two challenge is maintaining that culture of a small company as you're growing into a larger company.

HM: How do you help foster a culture of innovation at Zoom?

EY: It comes down to our values of caring for our customers. We listen closely to customer feedback and incorporate that customer feedback into new features and new use cases. Having open communication and being receptive to customer feedback is key. As long as we do that, we'll be fine.

HM: What are you most passionate about?

EY: What I'm most passionate about is making our customers happy, improving their communications, productivity, and culture.

Diving Deeper in the New CIO Mandate

The spectacular meteoric rise of customer-centric business strategies has transformed the role of tech leaders and elevated technology from the back office to center stage. The incredible success of companies such as Amazon, Netflix, Apple, Google, and Microsoft has proven conclusively that technology has the power to drive global markets and create value that would have been unimaginable in previous eras.

I spoke recently with Wendy Pfeiffer, the CIO at Nutanix, the global leader in hyperconverged infrastructure for cloud computing, and asked her for her perspective on the new

CIO mandate. Wendy had written an excellent guest column and delivered a fabulous presentation at the HMG Strategy Silicon Valley Global Innovation Summit in Menlo Park in February 2019, and I wanted to get her latest thinking on the transformational role of the CIO in rapidly changing times.[4]

"We're looking for the intersection of great technology and operational excellence. That's what IT is all about," Wendy says. "The CIO's job is marrying those two disciplines."

The best CIOs find the right balance, using superior technology to achieve and sustain operational excellence. "The modern enterprise needs transformational technology that is contained, framed and enabled by a robust operating model," she explains. That's why we need hybrid technologies and why we need to operate in a hybrid mode."

The 21st-century IT department is a "team of superheroes" combining the skills and talents of multiple individuals to reach a strategic objective. "We have visionary architects. We have operational experts. We have portfolio managers and support people. It's totally a team effort," Wendy explains.

For many organizations, however, operating in hybrid mode is genuinely challenging. Leading cross-functional teams in a modern global enterprise requires astounding levels of focus, discipline, and understanding. "The greatest challenge in IT is optimizing hybrid operations and managing diverse environments," she says.

The good news is that technology leaders are well equipped and perfectly positioned to guide the enterprise through the transformations that will be necessary for optimizing hybrid strategies. According to Wendy, "We have the skills and we have the experience."

I admire Wendy's confidence and her sense of purpose. She is a true technology leader and a role model for all of us. From my point of view, senior technology leaders have become absolutely indispensable to the modern enterprise. As Wendy notes, business now moves "at the speed of the machine." When you're moving ahead at full speed, there's really no margin for error—you need the perfect blend of technology and operational excellence to stay in the game.

Legendary Lessons from Andreessen

My good friend Jedidiah Yuch is the CEO and founder of Delphix, the amazing software company that provides a DataOps platform to connect, secure, virtualize, and manage data for many of the world's biggest companies. Jed participated in the HMG Strategy CIO and CISO Executive Leadership Alliance meeting in Menlo Park recently, and during our discussion on startup management, he spoke about legendary venture capitalist and serial innovator Marc Andreessen.

Jed cited Andreessen's genuinely groundbreaking 2007 post about what really matters most to startups, and I want to share portions of that epic post with you today.[5]

And so you start to wonder—what correlates the most to success—*team*, *product*, or *market*? Or, more bluntly, what causes success? And, for those of us who are students of startup failure—what's most dangerous: a bad team, a weak product, or a poor market?

Andreessen then offers his own definitions of *team*, *product*, and *market*. His thoughtful perspectives on these terms are incredibly useful. But he doesn't stop there!

If you ask entrepreneurs or VCs which of *team*, *product*, or *market* is most important, many will say *team*.... On the other hand, if you ask engineers, many will say *product*.

Then he offers his own opinion:

I'll assert that *market* is the most important factor in a startup's success or failure.... In a great market—a market with lots of real potential customers—the market *pulls* product out of the startup.

Moreover, Andreessen states:

The market needs to be fulfilled and the market *will* be fulfilled, by the first viable product that comes along. The product doesn't need to be great; it just has to basically work. And, the market doesn't care how good the team is, as long as the team can produce that viable product.

I find the clarity of Andreessen's vision and the precision of his language absolutely astonishing. Andreessen wrote the post in 2007, but it remains supremely relevant and useful today.

I'm glad that my friend reminded us of the value of Andreessen's words and insight. Marc Andreessen's unique status as a key figure in the success of Silicon Valley and the global tech culture is undisputed. A dozen years after he wrote the post, it still inspires us and provides essential lessons. Bravo, Marc Andreessen, and thank you, Jedidiah Yueh.

Critical Steps for Achieving Strategic Goals in the Modern Enterprise

I had an excellent conversation recently with Sheila Jordan, SVP and CIO at Symantec, the global leader in cybersecurity. Sheila is responsible for driving Symantec's IT strategy and operations ensuring that the company has the right talent, stays ahead of technology trends, while maximizing technology investments.

Sheila has a remarkably clear vision of the CIO's roadmap for 2025, and she genuinely understands the shifting landscape. In our conversation, I asked her to describe her top focus areas and key action steps for achieving the company's strategic goals.

"For CIOs, one of the first steps in digital transformation is getting rid of redundancy, duplication, and legacy applications that are breeding grounds for security issues," Sheila explains. "You need to eliminate the impediments preventing a smooth customer journey across your organization."

From Sheila's perspective, simplicity is essential to success in the modern enterprise. "You want to remove barriers and create frictionless journeys. That's how you extract value from your technology."

I love how Sheila places the primary focus on generating value for the business. "Operations are important, but extracting value and enabling key business processes are what make the organization great," she says. "As CIOs, we have a unique vantage point allowing us to see horizontally across the entire organization. We know where the gaps are, and we know how to stitch together multiple processes to create seamless customer experiences."

Exemplary CIOs focus on data first and applications second, she notes. "You need to shift your mind-set and focus on what's valuable to the organization. That's why you want a strong data taxonomy and clean house with the fewest impediments."

Sheila Jordan is also a strong proponent of planning, something that's become a lost art these days in many organizations. "Up-front planning is always beneficial," she says. "Some people think it slows them down, but in reality, planning helps you accelerate much more quickly and move faster. Having an end-to-end architectural view of both process and systems allows both IT and business professionals to visualize both upstream and downstream dependencies when creating this horizontal customer journey."

I agree completely with Sheila's advice. Executing against a plan enables you to move more rapidly and with more confidence. That's a key lesson for all technology leaders.

Business cycles have become so tight in today's markets that we often feel the need to move quickly, and sometimes we take shortcuts in planning. But even in our fastest-changing markets, that's rarely a good practice.

We constantly need to remind ourselves that planning is a critical step in the strategic process of digital transformation. I'm grateful to Sheila for generously sharing her insight and experience, and I sincerely look forward to my next conversation with her.

Meanwhile, I highly recommend reading her new book, *You Are NOT Ruining Your Kids: A Positive Perspective on the Working Mother.*[6] It's an excellent book, filled with timely tips and actionable advice for parents and their families coping with the realities of 21st-century living.

CIO as a Customer-Centric Leader

As enterprise companies prioritize their digital transformation initiatives, improving the customer experience has become a top mandate for CIOs.

It's hardly surprising, then, that top-performing CIOs spend more time with customers, focus on innovation, and do the work to enable core enterprise capabilities for both internal and external use, according to research conducted by Harvey Nash/KPMG and MIT's Center for Information Systems Research.[7]

"An important aspect of the CIO's role in cultivating a culture of customer centricity is to truly prioritize IT capabilities from a business perspective," says Steve Phillips, CIO at Alorica. "We are challenging the status quo to deliver innovative technologies that ensure world-class customer experiences at scale. As CIO, I want to assess the IT team's capabilities and performance through the eyes of our customers—I measure system availability based on whether my systems are always on. I measure project delivery based on my team delivering new capabilities in line with customer expectations. And I measure customer satisfaction by asking open-ended questions about how well, or otherwise, my team is doing."

On a granular level, customer-centric IT organizations seek feedback from frontline teams that are locked into moments of truth with customers, says Dan Roberts, president and CEO at Ouellette & Associates.

"Think about those moments we have with both internal and external customers," Dan says. "The best customer-centric organizations will create journey maps, focus on critical moments of truth and deliver seamless, frictionless experiences."

Dan points to Outback Steakhouse (owned by Bloomin' Brands) as an organization that has identified 60 moments of truth that arise with customers during their dining experiences. "If you can eliminate the friction points for any of these, such as the length of the wait time or the ability to pay your bill quickly so you don't feel like you're being held hostage, these are steps that can help bring customers back," Dan says.

There are lessons that CIOs and their companies can learn from other customer-obsessed organizations, such as Amazon. For instance, former SpaceX CIO Ken Venner points to how Amazon founder, chairman, and CEO Jeff Bezos keeps one chair unoccupied during meetings to remind attendees that the seat is for the customer, who may not be in the room but about whom everyone should be thinking.

Meanwhile, the metrics that are tracked to gauge performance should be taken from the customer's lens and focused on those things that customers would care about, adds Ken.

Another part of the CIO's role in fostering a culture of customer centricity involves engaging both IT team members and other employees throughout the enterprise on the company's mission. "An effective workforce strategy for customer-centric leaders must include finding ways to engage and empower people throughout the organization to play their part in delivering outstanding business results," Steve Phillips says.

"As CIOs and technology leaders who operate in an environment where technology is often thought of as the top priority, it's important not to lose sight of good leadership practices that can make all the difference." Key leadership practices that Ken cites include these:

- Provide clear line of sight for each employee on how he or she contributes to the success of the organization and its clients.

- Create an environment where individuals can be part of a team and influence operational excellence.

- Make available career development and progression opportunities.

- Ensure consistent, transparent, and fair leadership.

- Deliver timely performance feedback and fair compensation plans to all employees.

CIOs and IT teams also must ensure that they're applying customer-centric practices not only with external customers but also with company employees who are customers of the IT team, Ken Venner notes.

"Get out and talk with people and get a feeling for how well the IT team is engaged to help and support the company," he explains. This includes creating IT services feedback/survey methods to hear from both employees and external customers about exceptional service experiences as well as moments where the service could be improved.

"Create processes to jump on and remedy events where customer service has been below expectations," Ken recommends.

Given the speed of change and disruption in the industry, CIOs also need to create what Dan Roberts describes as a "culture of learning agility."

"The half-life of a skill is 18 months today," Dan says. "This suggests that, as a CIO, I need to create a culture of learning agility, where people are constantly learning, researching, studying and staying current on trends. People who do this are future-proofing their careers."

Another cultural aspect of fostering customer-centricity encompasses CIOs helping their teams to visualize and better anticipate where customer needs and expectations are heading. "Eighty percent of IT organizations are skating to where the puck was while roughly 20 percent are skating to where the puck is at the moment," Dan says. "It's just a handful of companies that have cracked the code on skating to where the puck is going to be."

Tech Leaders and the Customer Experience

When companies push further ahead with their digital strategies, it becomes increasingly evident how CIOs play an integral role in helping their companies deliver enhanced customer experiences.

The proof is in the results. According to the Harvey Nash/KPMG CIO Survey 2018, which canvassed 3,958 IT leaders, customer-centric organizations are 38 percent more likely to report greater profitability than those that aren't customer-centric.[8]

CIOs can step up their involvement in fashioning top-rate customer experiences in multiple ways.

"CIOs should start by ensuring that their own IT organizations are customer-centric," says Nathalie Rachline, former COO at Visual Farms. "If a CIO does not have their organization structured with their internal clients with customer centricity in mind, they won't be well positioned to help the business address the end customer."

"CIOs need to be change agents," says JP Batra, CTO and principal at Blue River International, Inc. "Technology-based disruption has become pervasive! Being technologists themselves, CIOs are best positioned to educate their peer CXOs on emerging tech's impact on business and suggest ways for the business to take a leadership role."

Many CIOs are also becoming more involved in fostering a culture of customer centricity within their organizations. One way to do this is by taking a deeper dive into customer data and sharing these insights with the CMO and other key stakeholders.

"CIOs should look at the data they have about customers— what they buy, how they buy, and how they use their products to identify patterns. This data can be used by product development and marketing teams to determine where their customers are going," says Helmut Oehring, executive vice president at Asteelflash.

It's also important to look at things from the customer's point of view. "You have to start thinking in terms of what customers value—shifting from a focus on products to those things that are value-driven for customers," JP says.

Until recently, CIOs have rarely been involved in a company's product development cycle. But by doing so, CIOs and their teams can better understand how customers are using a company's products and ascertain additional features and functionality that customers may be seeking, Helmut explains.

Of course, one of the best ways for CIOs and their teams to gain a deeper understanding of customers is by spending more time with them. "The CIO needs to be a part of those teams that meet regularly with customers," Nathalie says. "These types of interactions can help CIOs to better understand what customers are looking for."

One of the ways that CIOs can work with line-of-business leaders in delivering on customer goals is by utilizing technologies such as AI along with customer data and analytics to better understand and respond to customer needs. "This shifts CIOs from being order takers to a partner that's delivering value," says JP.

Guiding Structural Change Across the Enterprise

I spoke recently with my good friend Barbara Cooper, the former CIO of Toyota North America. Barbara is now a consultant and an executive coach in Phoenix, and she's always a wonderful source of insight.

From my perspective, Barbara is a genuine pioneer in many ways. She's had a highly successful corporate career spanning four decades and crossing five industries: retail, financial services, municipal government, logistics/distribution, and automotive. She is an innovator and risk taker known for her passion for business alignment, organizational strategy, and optimizing and developing exceptional IT talent.

Barbara became the first female vice president in the technology function at American Express and led the company globally into the personal computing and networking generation, connecting travel offices and operating centers around the world together for the first time. At Toyota, her reputation as an innovator elevated her to the top echelons of technology leadership.

I asked Barbara to describe the evolution of the CIO role, and her response was especially timely. "Today's CIO has to be comfortable with taking risks," she says. "Modern markets move with unbelievable speed. Keeping pace with them requires taking risks. In the past, CIOs didn't have to manage high levels of risk. Now they do."

I admire Barbara's succinct analysis of the difference between the traditional CIO and today's version of the role. "The modern CIO is more of a chief innovation officer and a chief digital officer," she explains. "These days CIOs are far more likely than their predecessors to be engaged in the customer experience and involved across the entire supply chain at every conceivable touchpoint."

The CIO's role has been elevated from a back-office job to a key business strategy role with major responsibilities spanning the enterprise. "For the modern CIO, everything is on the table. The digital age demands new business models that require new internal structures within the corporation," Barbara explains. "The CIO is perfectly positioned to guide the business through the structural changes that are necessary for getting the maximum value from technology investments."

Her comments are spot on. It's not enough to talk about being an "agile enterprise." Great companies empower their CIOs to make the structural changes necessary to make sure that information flows smoothly and seamlessly across silos and departments. From my vantage point, the ability to guide corporate restructuring is definitely part of the modern CIO's portfolio of executive skills.

Transforming a Major Industry

As the CIO at Perrigo Company plc, Thomas Farrington is keenly aware of the disruptive influences that are shaping the retail self-care industry. As PwC pointed out in a 2016 report,

"The Pharmacy of the Future: Hub of Personalized Health," plummeting reimbursements and industry consolidation are forcing pharmacies to look to new sources of revenue and redefine themselves.[9]

Indeed, consumers have been quick to adopt retail pharmacies as their neighborhood sources of flu shots and strep tests along with milk, makeup, and over-the-counter medications. Moreover, research reveals that consumers would be open to more services if retail pharmacies offered them.

Perrigo is an international manufacturer of branded and private-label self-care, over-the-counter pharmaceuticals. These include store-brand versions of cough, cold, and allergy medicines, analgesics, gastrointestinal products, smoking cessation products, and infant formulas.

I had an in-depth conversation with Thomas to discuss some of the ways that he and the IT team are helping the 131-year-old company to succeed in this fast-evolving industry. Here's a lightly edited version of our discussion:

Hunter Muller: Where is Perrigo in the current innovation cycle and where are you placing your bets over the next few years?

Thomas Farrington: When we look at the dynamics of retail and the competitive landscape, it has created a call to action to recognize that things have changed. When you think about the revolution in retail where we are reimagining and reinventing the retail digital experience, driving value from transactional data and business intelligence are center stage for companies for understanding markets and, specifically, the end consumer experience; and where do we need to double click on augmented intelligence? What we need to think about in the business are those technologies that are going to help us to succeed as a business today and going forward.

We're at an exciting point where there's a convergence of technologies and challenges that is creating immense opportunities for us as a business.

HM: What are some of the ways that artificial intelligence plays into this?

TF: When you look at the products that we sell in Europe as an example, we draw on customer insights from marketplaces where our products are represented. Understanding the customer experience through social listening and understanding marketplace reviews is key to staying relevant to our customers.

There is an enormous amount of industry data and internal transactional data that we have just begun to tap into for its value. As we become more agile in this space, AI will become a major player in freeing up company resources to focus on data analytics and transforming our views from "rearview mirror" reports to predictive analyses that shape strategic and tactical decisions ranging from what we manufacture across the entire product lifecycle to how we market to certain demographics.

HM: How would you characterize your role in the business and how you and the IT organization are viewed?

TF: In my first eight years with the company, we were involved in a substantial amount of inorganic and organic growth. My role over that period was ensuring we had an operating model that was scalable, well controlled, and operating at a "best cost" price point.

We were also looked to as a source for non-product-related innovation in the company. In our traditional role, we researched and implemented technologies ranging from automation of manufacturing processes on the shop floor with automated guided vehicles to decision support systems for inventory and supply-chain planning.

In a nontraditional role, we were an "innovation lab" for the business in the area of e-commerce. My team unpacked the DNA of the e-commerce business model to understand how marketplaces such as Amazon and Alibaba work and translating that into an architecture of processes, controls, organizational capabilities, and technologies needed to be relevant in manufacturing, marketing, and selling in a digital world.

HM: Tell us about the work you're doing with PwC on cybersecurity and governance.

TF: The life sciences industry had a wakeup call a few years ago when a very reputable company was hacked. When you step back and look at what happened, they weren't the intended target.

In the past year, more than 60 percent of my time has been spent looking at where we need to go from a cybersecurity perspective. I reached out to PwC to help us to better understand our points of vulnerability and how to break down those walls that exist to better understand our assets and the protection of those assets and what our maturity levels were.

We made real progress in the last year in these areas, in our maturity, our organizational capabilities, processes, and tools. With the help of external advisors, we formulated our transformation based on our business model and less on headlines in the media, prioritizing investments that are relevant to our business and our level of risk appetite as it applies to cyber.

While you are only as good as your next zero-day vulnerability, we have a comprehensive approach to our risks relative to the NIST framework and are making balanced investments in identifying, protecting, detecting, and responding to risks.

HM: How do you help make the business more nimble and responsive while addressing compliance requirements such as GDPR?

TF: The IT and Compliance teams partnered to work with our business units and functional groups to use GDPR as a rallying point not only to assess compliance but to take a long view on the value that the data collection and applications were returning over time. There were a number of legacy applications that really weren't needed or that, upon looking at the risk or effort to remain compliant, didn't provide a good return to the company.

By "weeding" the portfolio with a value lens, we not only achieved compliance by the required timeframes, we reduced the cost of compliance considerably. Going forward, we now have a strong process framework to ensure that we're not only compliant but that we are keeping better inventories within our portfolio and certifying the ROI meets company objectives for our investments. Also, if you're an external vendor that works with us, we've identified for them how they impact our GDPR compliance.

I've learned over my lifetime that you need to go slow to go fast. With GDPR, we did a great job of compliance leadership across the company and with marketing on what GDPR compliance constitutes.

HM: What are some other initiatives you're currently involved with?

TF: We've just made a commitment to go with SAP HANA. We've leaned into a platform that will enable us to actively govern and manage master data, and perform more real-time operations and predictive analytics. We believe the investment will be foundational in moving the value needle for the business.

HM: What's your relationship like with the CEO?

TF: Our current CEO is relatively new to the company, just under a month at the time of this interview. As a seasoned consumer products CEO, he brings a vision, expertise, and energy to use technology beyond the base transactional value that is essential to run daily operations. As his CIO, he looks to me to enable his vision and move our vast internal and external data to transform the way we do business—which takes trust, collaboration, capability, and execution.

We seem to be off to a great start thus far. Our CEO is a transformational leader and our biggest advocate for investment in strategy, data, and tools to drive shareholder value at the pace shareholders and end consumers demand. In short, we are set up for being able to accomplish great things for the company with great support from the corner office.

Riding the Artificial Intelligence Wave

Bhavin Shah is a talented, three-time entrepreneur, CEO, and founder of Moveworks. The company's purpose-built platform solves IT support issues instantly and automatically using AI.

We proudly partner with Moveworks, and have been astounded with the company's success and growth since we have incorporated them into HMG Ventures, a thrilling venture capital investment component expanding within our company.

In an in-depth discussion, Bhavin shared the most critical winning strategies by which his company garnered 10x value growth in just three years. Here is some of what he said:

In 2016, when we founded Moveworks, we set out to build an enterprise SaaS platform that had AI and machine learning at its core. Machine learning is so fundamental to what we do that if you took the machine learning out of Moveworks, you'd have nothing left—our product just wouldn't work.

In the five years before starting Moveworks, we witnessed many major leaps forward in the performance and application of AI in a number of different categories: from image recognition, language translation, and voice recognition to robotics, predictive analytics, and expert systems.

Most of the newsworthy advances were being made in the consumer space, by companies like Google, Facebook, and Amazon, because these companies had access to vast data sets and the seemingly limitless compute power required.

But we were interested in forging a new frontier for AI in the enterprise, where data sets are often small, messy, and siloed. We quickly homed in on our use case—autonomously resolving IT support issues submitted by employees—and this focus helped us to build an incredible solution that is transforming the IT support experience for employees around the world. Focus is a critical asset when working with bleeding-edge technology.

Today, AI is one of the hottest topics among IT leaders. Every CIO says they want it. Most vendors will tell you they are doing it. But what exactly is it? And how do you make the most out of it?

First, let's be clear on the terminology. AI is a very broad term that can describe any machine or system that exhibits intelligent behavior. "AI" is often used interchangeably with "machine learning," but they are not the same thing. There are several categories of AI that are making strong progress at the moment: image recognition, augmented reality, virtual reality, natural language processing (NLP), speech or sound recognition, conversational AI, natural language generation, voice generation, robotics, expert systems, predictive analytics, and so on.

You can think of machine learning as the math, algorithms, and models that make these systems work effectively. Traditionally software has been programmed using deterministic logic—a mesh of "if this, then that, else" type statements. The problem with this is that you have to program every piece of decision-making logic ahead of time. Instead, machine learning uses data to make decisions and can learn continuously as the data shifts. It is far more scalable and accurate, if you have sufficient training data.

At Moveworks we refer to ourselves as an AI company because to solve our use case, we had to combine multiple categories of AI—NLP, conversational AI, predictive analytics, expert systems, and so on. I think *AI* is the correct term to use in this scenario. Conversely, you'll see companies that have implemented a single machine learning model into their technology claiming they are now "AI." That's good for marketing, but it's an incorrect use of the term.

Having built an AI company from the ground up, my biggest observation is that AI is as much a cultural and philosophical shift as it is a technology shift. Andrew Ng famously described AI as "the new electricity," and I like this analogy on many levels.

Prior to the introduction of the electric dynamo, which kickstarted the first Industrial Revolution, factories were powered by steam engines. The energy produced by a single steam engine would rotate a giant axle that ran down the middle of a factory, and the individual machines would be connected to this axle by pulleys. All the machinery in the factory had to be physically located around this giant axle. And the axle was either on or off—you either powered the whole factory, or nothing. With all of these moving parts and steam engines powered by fossil fuels, factories were dangerous places to work.

When electricity entered the scene, most factories simply switched out their steam engine for an electric dynamo. Same giant axle, same pulleys, same machines, different energy source. The expected productivity gains were not realized. But over the next 20 years, factory owners started to build entirely different machines with individual power sources and mini-motors. This led to a complete redesign of factory floors, now untethered from the axle. This freedom led to the more efficient manufacturing processes pioneered by Henry Ford and then the Japanese Kaizen process. With more autonomy over their machines and work and cleaner, pollution-free work environments, the skills and salaries of factory workers increased.

Unlocking the true benefit of electricity required a huge cultural and philosophical shift in the way factories, machines, and processes were designed. This is important to remember.

We have witnessed a similar phenomenon with AI while building the Moveworks platform. Too many IT leaders are trying to wedge in some AI and machine learning to their existing processes and hoping that they will see huge benefits. The real opportunity is to completely rethink the way you work.

Designing and running electricity-powered factories required a different set of skills to steam-powered factories. Similarly, building AI requires a very different talent pool to traditional technologies: ML modelers, ML platform engineers, data scientists, data evaluators, hardware (GPU) specialists, and others. These people are well educated, highly skilled, and in high demand. So it's best not to treat AI as a side project. You have to go all in and build the right team and culture.

For this reason, my advice to IT leaders is to be very clear on when it makes sense to work with a vendor versus building something yourself. The simplest framework I can give you for this is to think about the data for your use case. If you are the only company in the world that has access to the data, you have to build it yourself. But if your use case is common across many companies, then a vendor that can build a platform leveraging these multiple disparate data sets is going to build more accurate models than you could on your own.

For example, at Moveworks, we realized that IT teams across the world were all solving the same IT support problems for their employees. So we pioneered a technique called Collective Learning that leverages all the support issues across all our customers to train our models. For this reason, no individual IT support organization could ever build machine learning models as accurate as ours. We now have a significant data advantage over others.

So a critical skill you need to develop is understanding whether a vendor really knows what they are doing with AI. With traditional technology, you might focus more on evaluating features and functions. But with AI, you're trying to evaluate the team: how quickly they can take the latest research and productionize it; how efficient they are at retraining models when they drift; what is their approach to annotating and labeling data; how they leverage data across companies.

If you're going to build your own AI, then you need someone on your team who is an expert in machine learning–based systems: someone who can experiment with the latest models and get them into production; who can attract high performers from a scarce talent pool; who is equally adept at talking about code, math, and hardware.

As a final thought on AI, you have to remember that we are still in the very early stages of figuring out what's possible. New models, architectures, and hardware are being released all the time. For example, when we founded Moveworks in 2016, we were using word embeddings—an NLU technique that leverages deep learning to understand the semantic meaning of words. But then in 2017, Google released a paper describing a new sentence embedding technique that could determine the semantic meaning of an entire sentence. We incorporated this into our platform and saw 10 to 15 percent performance gains in some of our models. In 2018 the best practice architecture for embeddings shifted from Recurrent Neural Networks (RNNs) to a new Transformer architecture. And then in

2019, we saw the introduction of several new NLU models, such as BERT. We're now seeing performance gains of up to 25 percent with the introduction of BERT.

This is the life of an AI company. We are constantly reviewing the latest research and finding unique ways to implement it in production. It is both science and art.

Riding the AI wave is incredibly exciting, and I think collectively we are more aware of how to leverage new technology than our counterparts were during the first Industrial Revolution. We can learn from their mistakes. But the task for all of you is to push the boundaries of your own thinking. AI is going to upend many of your existing processes, for the better. Just be ready to embrace the change.

Notes

1. Louis Columbus, "IDC Top 10 Predictions for Worldwide IT, 2019," *Forbes*, November 4, 2018.

2. IDG, "2019 State of the CIO," January 17, 2019, https://www.idg.com/tools-for-marketers/2019-state-of-the-cio/

3. Deloitte Insights, *Tech Trends 2018: The Symphonic Enterprise*, 2017, https://www2.deloitte.com/content/dam/insights/us/articles/Tech-Trends-2018/4109_TechTrends-2018_FINAL.pdf

4. Wendy Pfeiffer, "The New CIO Mandate," *HMG Strategy*, February 19, 2019, https://hmgstrategy.com/resource-center/articles/2019/02/19/the-new-cio-mandate

5. Marc Andreessen, "The Pmarca Guide to Startups, Part 4: The Only Thing that Matters," June 25, 2007, https://pmarchive.com/guide_to_startups_part4.html

6. Sheila B. Jordan, *You Are NOT Ruining Your Kids: A Positive Perspective on the Working Mother* (inCredible Messages Press, 2018).

7. Harvey Nash/KPMG CIO Survey 2018, "Special Report: MIT CISR," https://www.hnkpmgciosurvey.com/special-report-mit-cisr/

8. Harvey Nash/KPMG CIO Survey 2018, "The Transformational CIO," https://assets.kpmg/content/dam/kpmg/be/pdf/2018/07/CIO_Survey_2018_Harvey_Nash_report.pdf

9. Health Research Institute Spotlight, PwC, "The Pharmacy of the Future: Hub of Personalized Health," May 2016, https://www.pwc.com/us/en/industries/health-industries/library/pharmacy-of-the-future.html

Chapter 2

Leading in Disruptive Times

There are incredible opportunities for companies to leverage digital platforms and technologies for creating new business models and delivering differentiated customer experiences.

Organizations need bold and courageous leadership. They must provide employees and other stakeholders with a clear vision for what the future holds—and the journey the organization must take to get there.

"It's no longer enough for a CIO to lead an IT organization. Today's CIO must drive change management across the enterprise in order for the company to succeed," says Bob Concannon, senior client partner, CIO/IT Officers Practice at Korn Ferry. Bob was one of the distinguished speakers at

HMG Strategy's 2018 San Francisco CIO Executive Leadership Summit held in April.

As CIOs have stepped out of the back rooms and into the spotlight in helping companies to craft business strategy, they've also needed to disrupt themselves and be receptive to other opinions. According to Tony Leng, practice leader and OPM at Diversified Search, "To be a great leader, you need to be open to new ideas and other perspectives."

This includes willingness among CIOs to update their leadership style and be transparent in their communications with peers and employees. "How are you going to change your organization if you can't change yourself?" Tony asks. "You have to be authentic if you want people to listen and follow."

It's also essential for CIOs to take innovative approaches to problem solving. "Many of the things that make you good as a CIO—such as delivering programs and projects on budget, and tracking uptime KPIs—prevent you from being successful as a disruptive, transformational leader," says Ralph Loura, SVP and CIO at Lumentum. "As a visionary leader, you must learn to think like a CEO."

For CIOs to successfully help disrupt their organizations, they've got to constantly evaluate what differentiates business today versus in the future. "You must also be willing and able to reinvent yourself at these pivots," says Naresh Shanker, CTO at Xeros Corporation.

Becoming Boardroom Ready: Advice from Board-Level Tech Execs

As corporate boards of directors continue to recognize the role that technology plays in shaping organizational strategy—from creating customized customer experiences to helping the company to differentiate its products and services in the market—board members are increasingly looking to bring technology executives into the mix.

In fact, more than one-third of the next-generation board directors placed on S&P 500 boards in 2018 have backgrounds in the tech/telecommunications sector, according to the 2018 U.S. Spencer Stuart Board Index Highlights.[1]

But CIOs, CISOs, and other technology executives are not being recruited just for their tech savvy. "Boards certainly need technology expertise," says Sheila Jordan, CIO at Symantec, who meets with her own company's board and is also a board member at FactSet Research Systems, Inc. "Most companies, whether consumer-oriented or B2B-focused, are striving for a frictionless customer experience, and the way to deliver that requires the coordination of marketing, engineering, sales, and IT to deliver on the customer journey. IT is horizontally focused and systematic in its thinking. To put together this customer journey, technology executives are needed on boards to help connect all of the pieces."

Because technology is deeply embedded in the operations of companies across industries, many boards continue to struggle with how best to leverage technology for competitive advantage, says Patty Hatter, SVP of Global Customer Services at Palo Alto Networks and former CIO at McAfee, and public company board member at II-VI.

Patty, who will be joining Sheila Jordan on the Silicon Valley panel discussion, points to how boards constantly wrestle with how best to utilize technology to help identify how to open up new markets, create new business models, and drive operational efficiencies. "It's such an applicable skill to have on a board."

"Every organization is moving from 'castle and moat' (analog) to 'platform' (digital)," says Tony Leng, who moderated the boardroom-ready panel discussion. "It is a tough journey (enabled by IT), and the business needs executives who understand this journey—along with a Board team that complements them."

"It goes to the understanding of IT and business being as one," says Patrick Steele, chair of the CIO Advisory Board at Blumberg Capital who also spoke on the panel. Patrick, former CIO at Albertsons, is on the board of directors at Saucey, a liquor delivery company, and CommonSpirit Health (formerly Dignity Health). He says that as companies have become more technology-enabled over the past decade, "it's good to have to have people on the boards who understand how technology can be an enabler in moving the business forward. In my opinion, it's a business perspective strengthened by technology expertise."

For instance, when Patrick first joined the board of Dignity Health in January 2013, the board was interviewing executives with both technology and finance backgrounds for open positions. "The reason we ended up selecting two technology executives was for their business acumen," Patrick notes.

Understanding the Role

Members of a board of directors have a dual mandate: advising on the strategic direction of the company along with their involvement with governance and oversight, says Patrick Steele.

"The time is consequently ripe for CIOs who have experience in this digital transformation to use these skills to help other organizations navigate this path," Tony Leng explains. "It is an amazing time, but the skillset required at the board level is different than at the management level."

Boards also look for executives who are effective storytellers and strong collaborators but who know when to sit back and listen, Sheila Jordan adds. "If you have too many mavericks who don't work well with others, the board gets disrupted," says Patrick Steele.

One of the main reasons that CIOs and technology executives are increasingly being recruited by boards is due to their overarching view of how the various pieces fit together for the enterprise to execute on its strategy, Patty Hatter explains.

"You're not coming in at the board level because you have the most detailed insights into a particular technology," she says. "You're coming to the board in the context of how you think about that in terms of the broader business and the ability to relate to other board members. You've got to bring that higher-level view to the table."

A good starting point for tech executives who aspire to corporate board positions is to start in their own communities by serving on the advisory board of a nonprofit, a hospital, or a charitable organization. "It's a great way to learn how a board works together and the qualities that make a good board member," Patrick Steele says.

Sheila Jordan strongly recommends that technology executives who evaluate board opportunities commit only to those organizations or industries that they have a passion for. "Board commitment is way more than four meetings per year. When an issue arises, you might spend time with that board that takes away from your vacation or personal time, so you need to have a passion for the work you're doing," she explains. "You also want to make sure that the board is set up for healthy debate. Diverse thinking is critical. You want your experiences and opinions to be heard and that you can really add value. Never forget, as a CIO we are already living in the 'eye of the tornado.' Sharing that with other board members is invaluable."

Voice of the CIO in the Boardroom

As companies progressively rely on technology to lower costs, power new business models, and gain a competitive edge, corporate boards are increasingly recognizing the value of having CIOs and other technology executives participate on boards with them for their technology/business expertise.

Indeed, the percentage of public companies that have appointed technology-focused board members grew from 10 percent in 2012 to 17 percent in 2017, according to a study conducted by Deloitte.[2] The percentage of public companies with tech executives on their boards almost doubled to 32 percent for "high performers"—companies that have outperformed the Standard & Poor's 500 Index by 10 percent or more for the past three years.

For guidance on what it takes for CIOs to succeed at the board level, we recently turned to Ralph Loura, SVP and CIO at Lumentum. Ralph, who has held CxO roles at Rodan + Fields, Hewlett-Packard, Clorox, and Symbol Technologies, also serves as a technology advisor at REL Advisory, providing strategic advisory services to startups and mature companies in the technology, commerce, and consumer goods industries. Here's a lightly edited transcript of our conversation:

Hunter Muller: Why should the CIO be a leader in the boardroom?

Ralph Loura: CIOs are increasingly being sought out for corporate boards as more and more companies face digital transformations and born-digital competition. CIOs are not just being placed on boards for their technical knowledge; they're also being tapped for their broader understanding of how technology can be applied to tackle business challenges. They're valued not only for their technical acumen but also for the business viewpoint for the art of the possible.

HM: What advice do you offer to CIOs who are interested in landing a board position?

RL: Board members aren't managers, they are advisors. The role of board members is to ask appropriate questions and to help the C-suite to identify where there may be opportunities for improvement.

One of the most important qualities of a successful board member is to be candid. People bring their best face forward when they engage with boards, but they often soft-pedal tough questions. Boards are there to provide external perspectives for better options. You as a board member with technical experience should be able to help the C-suite think through their choices, but be careful not to play armchair CIO. The board is there to help the CIO, not to judge them.

The worst thing you could do is to assert your strategy and vision over theirs without due cause. This isn't a competition. It is a coaching and oversight role.

HM: Which business metrics resonate most with boards?

RL: It depends on the conversation. For cyberrisk, there are a few metrics that matter. Stay away from volume numbers—"we detected 125 attacks this month"—as they are meaningless. Instead, the board wants to understand relative maturity and operational efficiencies.

As for platform or operational KPIs, the board doesn't care about technical comparisons, such as the organization's percentage of cloud versus on-prem. For KPIs, you should be able to connect what you're doing to the things that boards care about—helping to drive growth or expanding into a new geography through the use of technology.

HM: In becoming boardroom-ready, where do CIOs need the most coaching?

RL: Far too many CIOs have stayed clearly in their swim lanes (technology and operations), and I think those CIOs that are more sought by boards have played a role in P&L, corporate strategy, or go-to-market activities. If your company doesn't afford you an opportunity to contribute in that way, then at least spend time to understand those areas and how your work contributes to the broader company and its outcomes.

HM: Can you point to any examples where you've worked with a board in educating them as to how technology can be leveraged to tackle a particular business challenge?

RL: In one case we had board members asking about how we could use machine learning or AI to deliver a new experience. What we did was to use real and practical examples within our business to highlight what we could do with these technologies. In some cases, it's going to result in years' worth of work to validate the application of technology, including testing. Many technologies are promising but are early in the value maturity cycle. Having a working prototype is an important early step, but getting adoption and the user experience right are key and those things take time and energy.

You've got to know the industry that you're in and the company's makeup to understand what could be disrupted within your company. Where is there automation that can be applied that could change the economics of markets? Don't think in terms of codifying existing processes, but how might operating models change all together if certain assumptions were changed?

As a CIO, you're rewarded for delivering on your budget, for being a conservative steward of resources and assets. Now we're asking CIOs to take risks and to think disruptively. We didn't breed that skill into tech leaders, we bred conservative governance. This helps suggest the rise of the chief digital officer, someone who is a bit more entrepreneurial.

So, for CIOs, it comes down to defining the capabilities the organization has and where there are gaps where I'm going to double down on the future; and using nontraditional methods to go after those opportunities.

HM: What types of skills or characteristics are boards looking for in prospective candidates?

RL:　Boards are looking for additional talent and perspectives, particularly in the tech area. A CIO needs to think in terms of helping the company to think about technology bright spots that can be applied as opposed to being on the tech committee or the audit committee because of their background. You need to earn your way there through reputation and networking.

To land a spot on a public board, begin mingling with public board members. Be knowledgeable, do your homework. Educate yourself on the role of the board member. And spend some time on private company boards.

Navigating Amid Continual Change

We're living in highly disruptive times. Massive geopolitical shifts and digital disruption are prompting corporate executives to focus on speed to market and speed to innovation in today's highly competitive marketplaces.

Heightened competitive pressures are also inspiring CIOs and business leaders to think, act, and lead differently in order to drive successful outcomes in the digital era.

"To help the business to succeed, the CIO must be the change agent in the C-suite," says Chris Colla, VP and CIO at B&G Foods, Inc.

CIOs also must demonstrate courageous leadership in working with fellow members of the C-suite to drive transformational change. "If your executive team is focused on the next quarterly earnings call, are they likely to take risks and innovate?" Milos Topic, VP and CIO at St. Peter's University, explains.

There are a number of actionable steps that CIOs can take to drive innovation and help the organization gain a competitive edge. "CIOs have two things they need to focus on to enable innovation: the integration of IT with the business and the professional development of the IT staff," says Justin Lahullier, CIO at Delta Dental of New Jersey and Connecticut.

"To lead in disruptive times, we must embrace the values of the emerging workforce," Vipul Nagrath, Global CIO at ADP, explains.

It's also part of the CIO's role to inspire IT staffers. Vipul adds, "I tell my IT team members that they must have an emotional attachment to their work in order to generate the outcomes we're looking to achieve."

It's imperative for CIOs to obtain a variety of viewpoints to help identify and act on new business opportunities. "CIOs need to look both inside and outside the organization to gain different perspectives for driving innovation," says Hugo Fueglein, managing director and CIO/IT Practice at Diversified Search.

Long gone are the days when the CIO's primary role was to simply help "keep the lights on." Thanks to the value they demonstrate in identifying opportunities for improving organizational efficiencies and for leveraging technology to attack new business opportunities, CIOs today are now more widely viewed as trusted business partners to the CEO and as key strategic contributors to the executive team.

This helps explain why four out of five CIOs believe their role has increased in importance over the last five years and why their ability to contribute to corporate strategy is now deemed as their most important skill, according to 2018 research conducted by Forbes Insights with Intel and VMware.[3]

"One of the main roles for the CIO today is to continue to build awareness and education around opportunities that the CEO, CFO, and other CXOs may not be aware of," explained Jamie Cutler, SVP and CIO at Air Methods Corporation, a nationwide emergency air medical transportation company. "One of the main aspects of a CIO's job now is to share opportunities with the C-suite for applying technology to drive market results, improve competitive positioning, and generate cost savings."

Today's CIOs also must demonstrate fearless leadership in clearly communicating to peers which initiatives are viable and which ones aren't.

According to Matt Mehlbrech, VP, IT at CoorsTek, Inc., "The CIO needs to be a trusted advisor to the executive team. Because, at the end of the day, virtually all process improvements or initiatives involve IT these days. The CIO must guide and steer what makes the most sense and make execution a reality."

Meanwhile, another way to gain credibility with the executive team is by streamlining operations to free up IT resources to devote more time to value-added activities.

For instance, when David Bessen stepped in as CIO at Arapahoe County Government six-plus years ago, the IT organization was in the early stages of heading down the virtualization path.

"Now we're 90 percent-plus virtualized," David says. "Making this transition has helped us to reduce day-to-day maintenance work, which freed up staff to focus on other things. Creating this stability has helped to establish the credibility of our IT organization and to gain trust. Now we can pursue innovation opportunities because we're not putting out fires all the time."

As Bart Waress, vice president of IT at Discovery Natural Resources, sees it, the CIO should act as the harmonizer within the executive team.

"The CIO sees all aspects of the organization and should share that vision and communicate the opportunities to the executive team," Bart says. "We should be the great unifier, not pointing out what is wrong but what is possible. And where we can add value to those opportunities is really important."

Motivating and Inspiring the Enterprise

As a captain in the U.S. Marine Corps for nine years, Tom Peck learned several valuable leadership lessons that he's been able to apply throughout his distinguished career as a technology leader at GE, NBC Universal, MGM MIRAGE, Levi Strauss & Co., AECOM, and currently as EVP, chief information and digital officer at Ingram Micro. These include:

- Motivating and inspiring people to do things they may not think they're capable of doing.

- Continuously reminding colleagues of the importance of teamwork; teams win and lose together, and they're only as strong as the weakest link.

- "Officers eat last"—take care of the troops (employees), win their hearts and minds, and get them to want to climb that mountain for their company and for you as their leader.

- Connect with staff—show humility and empathy. Help them understand the mission, and give them the tools to succeed.

- Small unit leadership—just as in the military, apply small units to initiatives and empower them to be successful.

Tom applies his leadership skills across several different areas of the company. As B2B customers have come to expect frictionless B2C experiences, more and more companies like Ingram Micro are competing against the "Amazon effect," and end customers want to consume technology differently—not just via the cloud, but truly on-demand and anytime, anywhere, and on any device.[4]

I had a genuinely enlightening conversation with Tom recently, and we discussed his role in helping to take the world's largest technology distributor and supply-chain company to the next level. Here's a lightly edited transcript of our conversation:

Hunter Muller: Can you talk about your leadership style in working with your CEO, board, and the executive team on strategy?

Tom Peck: Working with our C-suite begins with speaking their language and talking about business outcomes, not about IT or project lists.

Because technology is so pervasive and critical to all companies across industries, it's highly important to be actively engaged in the strategies of the business and not wait to be called upon. As part of these discussions, technology should be positioned as an enabler, not as a tool, product, system, or cost.

Ultimately, small wins, and transparency help to build credibility and strengthen trust with my fellow members in our C-suite. Victories also allow us as CIOs and CDOs to fail periodically but maintain the confidence of our leaders based on built-up credibility and goodwill.

HM: What are the key focus areas for Ingram Micro these days?

TP: The industry is at an inflection point. With so much emphasis now on digital relationships and customer experience, we're seeing B2B shifting more toward a B2C customer experience. We've always known what research shows: Companies that are customer-focused grow faster. Today it's no longer about big companies eating smaller companies—now fast companies are eating slow companies.

So, our industry—and our company—is literally at a crossroads. We can no longer be only a distributor or a supply-chain company, regardless of how big we are or the role we play in the channel. In order to survive and succeed, we need to differentiate.

One of the ways we're doing this is by increasing our focus on our "as-a-service" capabilities. This includes helping our channel partners distribute third-party cloud solutions under a new cloud platform to accelerate digital commerce—CloudBlue—which we announced in the spring of 2018. We are looking to replicate this success with our recently launched IoT marketplace.

Meanwhile, as part of our push toward delivering more of a B2C-type customer experience, we're focused on providing our customers with frictionless channels. These include digital self-service experiences that are personalized and interactive. Clients now expect to buy at odd hours, receive instant answers to questions, and receive same-day delivery.

We're also rolling out new technologies and analytics to assist our partners and resellers in understanding what people are shopping for and, via product association and data, assist in recommending additional products, services, and even financing. Data will also assist in auto-renewals.

HM: What's the current state of the company's digital strategy? What are you focused on now as both the CIO and chief digital officer?

TP: As the CIO, I'm focused on more traditional technology delivery. As the CDO, my focus is around customer experience, including easing and accelerating adoption, while also helping our executive team identify and execute on new business opportunities.

From a customer experience standpoint, a lot of this is centered around the hyperconnected experience—anywhere, anytime, using any device—and identifying opportunities to monetize these behaviors.

At Ingram Micro, we firmly believe that those companies which master the interplay between products, services and technology, processes, and human relationships will win in their industries. Here's how:

The key requirement is a move from being "customer-aware" to "customer-led." This entails going beyond the "what" and "how" of customer behavior to reveal the "why" in understanding your customers' needs and motivations.

In addition, customers expect a seamless and coherent experience as they engage with a company and traverse channels, contact points, and systems.

To that end, we are focused on building digital experience platforms to unify marketing, commerce, and customer service across our entire customer lifecycle.

Our strategy is built on five pillars ranging from customer experience to secure and resilient infrastructure. By investing in and building competency in each pillar collectively, we can build solutions and deliver customer value through programs and projects.

We have rolled out Agile scrum methodologies in newly formed product engineering teams. We have global centers of excellence which deliver the technical depth needed while business unit CIOs help prioritize the top efforts that deliver maximum customer value and/or monetize value for the company—such as new online marketplaces, one invoice/one credit line/one experience, services billing/bundling, via as-a-service offerings, customer portals based on SSO personas, endless aisle experiences, and much more.

HM: What are some of the strengths you carry over from your time at AECOM, Levi Strauss, and MGM that are helping you contribute to Ingram Micro's customer strategy?

TP: Some of my key takeaways from past experiences include learning from failures from difficult situations with people, technology, processes, decisions and then taking steps to prevent those failures from occurring again.

It's also critical to leverage your partner ecosystem and supply base, including vendor partners, universities, and other third parties to assist in transforming your business.

With each of the previous companies I worked for, we had different types of customers—AECOM (governments, sports teams, private corporations), Levi Strauss (consumer, shopper), MGM (leisure traveler, convention traveler). From those experiences, I've come to appreciate the importance of understanding your customer—whoever they may be—watching them and immersing yourself in their experience.

It's also essential to know your competitors. If you don't differentiate your brand and the customer experience, you will be disintermediated. Have an external lens; don't be too inward-facing.

Finally, know when and how to innovate and, conversely, when not to innovate.

Connecting the Dots for E-Commerce

Thanks to their unique perspective as to how people, processes, and technology can be interwoven, CIOs are in prime position to help orchestrate innovation initiatives.

At Conair/Cuisinart, Global CIO Jon Harding is playing a key role in the company's e-commerce digital transformation initiative, which is aimed at providing the company with an additional channel to distribute its products for incremental revenue growth and consumer engagement.

As part of these efforts, which began three years ago in the United Kingdom and Canada and were expanded to the

United States last year, Jon has become coleader of the company's e-commerce initiative, championing the use of a standard e-commerce platform and digital marketing solutions globally.

"Looking at this from a global perspective, we're looking to sell items that big retail clients don't want to stock, like spare parts for Cuisinart appliances," Jon says. "We believe that by offering these items directly to consumers through e-commerce, we can strengthen consumer engagement. This in turn contributes to our goal of improving all aspects of customer experience (c/x) for the end consumers of our products."

To help execute on the e-commerce initiative as the company's global CIO, Jon is working closely with the various business unit leaders and their digital marketing teams within this global company.

"It's really a program of work across various geographies and within each geography there may be multiple projects going on," Jon explains. "We're using a common platform on the back end, and so there's a lot of coordination needed, and that's one of the strengths of our global IT team in executing on project management. We have a weekly call with the development partners covering all development activities across all lines of business. We can bring that discipline from the IT side while Marketing handles the creative side of developing the website 'look and feel' for each particular market."

Renewing the IT Value Proposition

The long-standing industry joke has been that the acronym "CIO" stands for "career is over." But as technology has become pervasive in helping companies in every industry to reduce costs, improve the customer experience, and provide opportunities to help the organization to gain a competitive edge, many CIOs have rightfully earned a seat at the table with CEOs and fellow members of the C-suite for their role in contributing to organizational strategy and execution.

This helps explain why more than half of CIOs (57 percent) report that the business expects them to assist in business innovation and in developing new products and services, according to Deloitte's 2016–2017 Global CIO Survey.[5]

To shed light on the attributes CIOs need to succeed in the fast-paced and highly disruptive modern enterprise, I recently spoke with two respected search executives, Shawn Banerji, managing partner, Technology Digital and Data Leaders Practice at Caldwell Partners, and Mark Polansky, senior partner, Technology Officers Practice at Korn Ferry. Here is a lightly edited transcript of our conversation:

Hunter Muller: What should be the CIO's role in helping to identify opportunities to transform the business?

Shawn Banerji: What we're hearing in conversations we're having with boards and various stakeholders is this battle cry of "IT Is Dead: Long Live Technology." Meaning that the traditional IT department is dead but companies are very much focused on leveraging technology to help drive operational efficiencies and to gain a competitive edge.

Mark Polansky: The CIO must be capable of recommending innovative ideas to be evaluated and tested in pilot programs . . . ideas that can deliver operational or competitive improvements to the business. A CIO might say to a business or functional leader, "If we did this, would it make your life easier or better?" Also, spending time with revenue-generating colleagues can typically help CIOs observe and perceive ideas and suggestions for technology-enabled and technology-driven business opportunities to be explored.

HM: Should CIOs play a more proactive role in communicating these opportunities to the CEO and the executive team?

MP: I'm taking your question literally, and I think it's important to say that if I'm a CIO and I take a value-generating idea to my CEO, and by doing so I go around my revenue-generating colleague, that's simply bad form. CIOs should bring opportunities to the attention of the appropriate business leader in a highly collaborative manner. Of course the CIO should be proactive, engage the right people with their innovative proposals and inspirations, and present these collaboratively.

SB: Absolutely. These days it's probably the single most important contribution they can make that demonstrates how they're delivering value beyond operating efficiency. Look at what AWS is doing as a model.

HM: What are some ways in which the CIO can help foster a culture of customer centricity?

SB: If the company wants to be externally customer focused, they need to develop a top-tier internal customer focus. Drawing from what IT teams do to delight employees and users, we can apply the same philosophies to delighting external customers. If you make it easier for people to do their jobs, they'll make it easier for customers to do business with the company.

MP: I like to see CIOs proactively and literally get out and speak with their company's customers, hand in hand with either or both a business leader and a marketing leader. In this way CIOs will discern what the customer is doing and thinking. And it gives CIOs the opportunity to observe and discuss customers and their processes in real time along with their colleagues, and even solicit customers' reactions to some innovative ideas.

Notes

1. 2018 U.S. Spencer Stuart Board Index Highlights, https://www.spencerstuart.com/-/media/2018/october/ssbi2018_summary.pdf

2. Khalid Kark, Caroline Brown, and Jason Lewris, "Bridging the Boardroom's Technology Gap," Deloitte Insights, June 29, 2017, https://www2.deloitte.com/

us/en/insights/focus/cio-insider-business-insights/
bridging-boardroom-technology-gap.html

3. "Role of CIO Is Changing and Growing in Importance, Say New Forbes Insights Studies," Press
Release, March 28, 2018, https://www.forbes.com/
sites/forbespr/2018/03/28/role-of-cio-is-changing-
and-growing-in-importance-say-new-forbes-insights-
studies/#27164441426c

4. For a definition of the "Amazon effect," see https://
whatis.techtarget.com/definition/Amazon-effect

5. Deloitte University Press, "Navigating Legacy: Charting
the Course to Business Value 2016–2017 Global CIO
Survey," 2016, https://www2.deloitte.com/content/
dam/insights/us/articles/3591_2016-2017-CIO-survey/
DUP_2016-2017-CIO-survey.pdf

Chapter 3

Achieving Future State Goals

As I've said and written many times before, there's no better time than the present to be a CIO. Part of the excitement is fueled by the opportunities CIOs have before them for using advanced technologies such as artificial intelligence, machine learning, blockchain, and analytics to help propel the business forward. But it also reflects the evolution of the CIO role in working with the CEO and fellow members of the C-suite to transform the business.[1]

As Comerica CIO and EVP, Sangy Vatsa is helping the bank move forward in the digital age. He's focused on three specific modes of its evolving technology strategy, which is called Digital2025: keeping the business operating, working in partnership with the business to transform the bank, and constructively disrupting the business. "If you don't disrupt your business, someone else will," says Sangy.

As Comerica has embarked on its business transformation journey, key areas that Sangy and his team are focused on include how best to optimize costs while delighting the bank's colleagues and customers to improve loyalty and create the most sustainable engine for its integrated revenue stream.

"In this regard, the technology organization becomes a co-creator and a business partner to other business teams within the bank. A business domain-savvy technology leader and a technology-savvy business leader collaborate to co-shape customer-centric products and services based on shared goals. This is the core theme of our two-in-a-box product development approach that we established two years ago," Sangy explains.

In support of these efforts, Comerica's technology has incorporated lean, design thinking, agile, and DevSecOps methodologies to drive transformational change at the speed of business.

Recent examples in the co-creation model at Comerica include applications within its commercial banking business. "We've looked at the terms of treasury services, which used to be an annual approach for getting projects funded and multiple years to deliver new services," Sangy says. "We can now deliver new releases in three months using lean and agile methodologies. The business and technology partners co-shape these releases with active engagement with the treasury customers. We can also fail fast if needed and learn fast from it for future opportunities."

Meanwhile, Comerica's use of DevOps (and more recently DevSecOps) practices has delivered numerous speed, security, and productivity benefits. "Under the new IT operating model, security is a key embedded feature of all digital capabilities that we are creating. In fact, embedded security and compliance are the core part of our technology guiding principle for the bank, which was established over two years ago," says Sangy. "We've recognized that security needs to be a built-in construct and not a bolt-on construct."

Looking ahead, Sangy sees blockchain and quantum computing as opportunities to constructively disrupt the business with the potential to better serve Comerica's customers and colleagues and drive higher efficiency and growth, including potential use cases that are being explored and experimented across the bank's multiple lines of business. "If we were to remove unneeded intermediary steps in business transactions and introduce frictionless intelligent automation in the value chain, we can reduce the costs for our customers and business while significantly elevating the colleague and customer experience," he says.

Scaling Innovation at the Speed of Business

There are many challenges associated with innovation, including convincing employees they won't be punished if innovation initiatives fail as well as other cultural aspects.

For enterprise companies that have hundreds or thousands of B2C or B2B customers, the ability to innovate at scale is another thorny challenge that must be met. According to a recent Accenture study, just 22 percent of C-level executives say they've found highly effective ways to scale their digital innovations.

One enterprise company that has repeatedly excelled in this regard is ADP. The Fortune 500 provider of payroll and HR services was the first company to outsource payroll at scale and has continued to deliver numerous payroll and HR innovations at scale for businesses of all sizes.

I spoke recently with Vipul Nagrath, global CIO at ADP, to learn more about how the company is able to innovate at scale and where its innovation journey is heading.

Hunter Muller: How do you approach innovation at ADP and communicate this to employees?

Vipul Nagrath: We don't have unlimted time or resources but within these constraints we tell people, "Here is your sandbox and go create." That's step one. Step two, to paraphrase Andy Jassy (CEO of Amazon Web Services) at AWS re:Invent last year, it's not just about innovating but being able to execute at scale.

Last year we processed over 68 million tax forms. That's scale. And that has to be understood, to build technology to be resilient and to do it at scale.

HM: Tell us about your journey from being a CIO to running a P&L.

VN: At the end of the day, it's about how technology translates to the bottom line. If you look around today, the most valuable companies are technology companies.

The technology leader today has to achieve more balance today than in the past. Technology has become so pervasive you can't escape it. You have to be able to apply it into the business. Ten years ago you could just stay in your lane as a technology leader. Today technology *is* the business. And you have to be able to deliver technology by a certain date to meet business requirements.

For our company, the product I'm now overseeing is our single largest revenue-generating platform.

HM: Who inspires you?

VN: There are many things I've admired about Steve Jobs, from his focus to his discipline to his pride. That lesson in pride—pride of authorship—things will follow from that. Another would be Elon Musk. He is just phenomenal at thinking of the wildest crazy ideas and figuring out how to move forward with them.

HM: Can you point to an example of an innovation initiative that was on the fence and how you were able to stick it out?

VN: There are a number of product offerings that we're going public with now that didn't just materialize overnight, they took a number of years to materialize. We had periods of trial and error before they finally came to fruition. Identifying those setbacks allowed us to make refinements and deliver on the promise of those offerings.

It's how our next-generation technology is headed and being able to deliver it at scale. Once it's available, it's not two clients that will want it, it will be thousands of clients with tens of thousands or hundreds of thousands of employees.

As a company, there are a number of firsts we're very proud of. This includes the outsourcing of payroll, which was a first for the industry.

We as a company are steeped in this innovation that continuously shows up.

HM: Did you anticipate that ADP would develop a reputation as one of the world's leading data science companies?

VN: I didn't when I first walked in here, but it became very clear shortly after I started meeting with the key players. We joke around, but "data" really is our middle name. What we've done at ADP is tap into our unprecedented data and couple with safe and ethical AI to create an actionable data platform. This platform allows business leaders and HR managers to make informed and efficient decisions that directly impact the bottom line. It is a key part of our technology and innovation story, and we're infusing it into every product we develop.

HM: Can you describe your go-to-market vision?

VN: At a high level, thematically, there are five themes that we're following. It's our focus on those five themes that make a difference:

1. **The Evolution of Work.** This includes the emergence of Agile organizations, the growth of the gig economy, and the emergence of dynamic teams.

2. **The Evolution of Pay.** More workers are looking to be paid in real time or on tailored schedules as opposed to the traditional pay cycle their company may offer. If a company offers to pay you every night versus every two weeks, where are you doing to go?

3. **The Evolution of Business.** We are seeing the globalization of the workforce, tightening labor, greater regulatory pressures, including privacy, that we have to worry about.

4. **The Evolution of HR.** HR is becoming more analytical as they continue to find ways to leverage AI and ML while increasing their focus on talent and engagement.

5. **The Evolution of Technology.** HCM technology is transforming into platforms plus app eccosystems, all in the public cloud.

HM: What are the key takeaways?

VN: • It's an amazing time to be a CIO and technology executive, to help lead, reimagine, and reinvent the company and have a material impact on the business.

- CIOs and technology executives also must develop a deep understanding of the business to be able to deliver value on the organization's go-to-market strategies.

- CIOs and technology executives also must harness the power of customer data and analytics to thoroughly understand and respond to changing customer behaviors and needs ahead of the market to gain a competitive edge.

Great Executives Leverage Deep Business Experience to Drive Lasting Transformational Change

One of the many aspects of my role that I truly enjoy is having fascinating in-depth conversations with brilliant thought leaders. I spoke recently with Angela Yochem, executive vice president and chief digital officer for Novant Health, a super-regional health-care system with one of the largest medical groups in the United States. Angela and her teams deliver the world-class consumer capabilities, differentiating technologies, and advanced clinical solutions that allow the high-growth system to provide remarkable patient care.

In our conversation, we talked about the deep and powerful impact of technology on our daily lives, especially in the area of health care. Thanks largely to advances in technology, the average life span has increased. In addition to living longer, we're staying healthy and active—again, mostly due to astonishing leaps forward in multiple technologies.

"When you think about how advances in technology have accelerated the evolution of medicine, it's simply amazing," says Angela. "We know that we must have a world-class tech capability to deliver the highest-quality care when and where it's needed. We move very quickly to save lives."

Advanced technologies enable health-care systems like Novant to greatly accelerate the speed of diagnosis and treatment, resulting in improved outcomes and helping patients recover more quickly and more fully.

Angela herself is a truly wonderful role model for modern technology executives, and she draws on a wealth of varied experiences that make her particularly valuable in a rapidly changing field like health care. She has served as EVP/CIO at Rent-A-Center, global CIO at BDP International, global CTO at AstraZeneca, and divisional CIO at Dell. She's held tech exec roles at Bank of America and SunTrust and held senior technology roles at UPS and IBM. In these roles, she built B2B digital product lines, grew digital retail channels (B2C), created technical services lines of business, and transformed global technology capabilities.

From my perspective, executives such as Angela bring unique levels of depth and insight to the role. In a very real sense, she is a Renaissance executive, prepared for any and all contingencies. Her comprehensive business background enables her to communicate smoothly and effectively across all levels of the modern enterprise, including the corporate boardroom and C-suite. Those executive leadership skills are incredibly important when making the business case for investments that drive lasting transformational change.

Focusing First and Foremost on the Customer Experience

Tom Keiser is a business and technology leader with over 30 years of global business transformation experience. He is currently COO at Zendesk and is responsible for global operations for the company.

Before joining Zendesk, Tom held CIO positions at two of the largest apparel specialty retailers in North America: Gap, Inc. and L Brands. Earlier in his career, Tom spent 12 years in management consulting with Ernst & Young and Capgemini. He has a long track record of delivering business results, building scalable business and technology operations, and increasing organizational speed of value delivery.

I spoke recently with Tom, and he shared his valuable insight with our team at HMG Strategy.

Hunter Muller: How is Zendesk continuing to differentiate itself in the market?

Tom Keiser: At our core, Zendesk still has the same focus, which is on the customer and on the customer experience. Zendesk is 13 years old and was formed to simplify and improve the customer service experience, both for agents and for customers. Our differentiator has always been about ease of use and speed of implementation and integration. We were built as a cloud and SaaS product from the beginning, with a focus on allowing customers to easily sign up, set up, and operate their business.

As we've grown to an enterprise software company, we've kept that "beautifully simple" mind-set. We've been on a product journey from customer support to modern CRM, and each year we add capabilities that bring the full customer experience and communication together—over the last two years adding machine learning, SFA, customer communication, evented customer data stores, and fully migrating onto AWS.

HM: Tell us more about Zendesk's machine learning capabilities.

TK: We initially focused on two interactions: The first is the customer's interaction with the company they're doing business with, serving up potential answers to their questions, initially on the email channel and now across all communication channels. The second was identifying and serving up gaps in content on the help center back to the customer service organization, so that they could add content to answer customers' questions more quickly.

Our focus on machine learning is to better enable customers to quickly get resolution to their questions and needs and to better arm customer service reps with information that will help them quickly resolve a customer's problem.

HM: Can you point to some of the ways that Zendesk collects and acts upon customer feedback?

TK: We get tons of customer feedback, as we now have over 100,000 customers. We gather a lot of usage data, which we use to constantly improve the customer experience. We are a customer experience company with a dedicated customer experience team: a 500-person organization—350 in our customer advocacy organization and 100+ in customer success. We get to see a lot of interesting use cases, not just traditional customer support and customer communications. We monitor our product, we're a multi-tenant SaaS company, and we can see where processes start and stop and where there are opportunities to improve the use of our product.

We also have grown up with nearly every disruptive company that has put the customer and customer experience at the center of their business model. Many, if not most, of those companies use Zendesk at the core of their customer experience and are now making the transition from disrupter to publicly traded company. The learnings from scaling and growing with these unicorns has been invaluable to our product and our company.

As Zendesk grows, we also grow into new roles. We've just hired our first chief customer officer, Elisabeth Zornes, who joined us from Microsoft.

HM: As Zendesk continues to move upmarket into more complex and larger enterprises, what are some of the differences between the needs of these customers compared to SMBs?

TK: It's been a journey. At our core, we're still very much a horizontal customer experience product. We want our product to be frictionless for customers to buy and use. The move upmarket has forced us to mature and evolve into a series of holistic customer services.

By this I mean anything from robust pre-sales demo services to professional services to help advise our customers through pilots and implementations, to account management and customer success to help them quantify the value of the evolving and continually maturing of their customer experience, to deep security compliance, guidance, and question answering. We've had to mature the processes that an enterprise company needs in place, and we've had to do it quickly. We've brought in experienced leaders to help us move as swiftly as possible. It's been incredibly interesting and exciting: really, the greatest show on Earth. We're evolving our selling from transactional to solutions-oriented. We're early in that journey. It's really changed the discussion with our customers about the full suite of solutions we offer.

HM: Can you offer some recommendations for CIOs to partner closely with line-of-business leaders and the CEO and the board in focusing on customer-centric strategies? Are there certain behaviors or tactics that tend to resonate well?

TK: It's a big deal. We had a panel at our user conference this past fall, and we talked about the evolution of customer experience with four great CIOs. The reality is that CIOs need to start with the customer, their company's customer, to architect their company's technologies and processes. For too long, CIOs have focused on the internal customer and build suboptimal processes and technologies. We're onboarding our CIO, Colleen Berube, to learn our customers' experiences so they inform what our IT strategies and approaches should be. Having that empathy and understanding of what works for customers is a big part of the CIO's role.

HM: To that end, can you also offer recommendations for CIOs to communicate effectively with the CEO and board of directors?

TK: As a CIO, you don't get to start with trust. You have to build a trusting relationship with the CEO and the board, beginning with honest and clear insights and then delivering on them. If this is pressured, forced, or rushed, it can damage the relationship with the CEO and the board. You have to invest the time with them.

The CIO role is a problem-solving role: We have to deal with problems. The key is to deal with, not just identify, them. The CIO has to be able to provide quick and clear context, and to have thoughtful and clear approaches to delivering bad and good news to the CEO and business leaders. CIOs have to keep delivering value and insights to remain relevant. Your ability to communicate clearly and concisely as a CIO is critical to your success.

HM: What are some steps that the CIO can take to help the enterprise achieve its future-state goals?

TK: We see a lot of focus on agility. CEOs are focused on their businesses being disrupted—even those businesses that are disruptors—along with a continual improvement of customer experience to fend off that disruption. So for CIOs, it's critical to look at their architecture and application layer and take advantage of the public cloud to remove the boat anchor that are legacy systems so that your business has agility and flexibility. It's about architecting to take advantage of SaaS—that's where the innovation is happening. If you can work your way into this service architecture, it puts your IT organization in a much more valuable position.

We're in a time where you can rapidly build out advanced analytics. But you have to do it thoughtfully, carefully, and aggressively to achieve value. These things are a paradox and have to coexist: This is why the job is so compelling and nuanced.

An interesting breakthrough that all CIOs have to be thinking about is that with 5G coming at us, the whole world is going to be reconnected with almost unlimited bandwidth. It's going to take the Internet of Things and interacting with devices to a whole new place. It's impossible to know what that's going to look like, but it should be part of your architectural planning for internal and external use cases.

It's the most exciting time ever to be a CIO, to be able to positively impact your business by building speed and agility into your business model, to be able to quickly add value, to help your business think differently, and to positively impact your company's productivity.

Enabling the Enterprise to Evolve

As CEOs look to expand their organization's growth in existing and new markets, the CIO is playing an increasingly important role in helping the enterprise to meet its objectives.

For instance, 40 percent of CIOs say their function will be vital to successfully developing customer-facing solutions, creating global capabilities, crafting new revenue opportunities, and fostering innovation within their companies, according to a 2018 study of 400-plus CIOs conducted by Forbes Insights with Intel and VMware.[2]

Clearly, one of the ways that CIOs and technology executives can help organizations in this regard is by identifying technologies that can help move the needle for the company and to potentially offer a competitive edge.

At Deloitte, the use of robotic process automation (RPA) technologies has resulted in substantial value for Deloitte and its clients across a number of applications, says Quintin McGrath, senior managing director of the Technology Management & Enablement group at Deloitte Global.

"Deloitte* is seeing significant benefits from RPA, not only from an IT perspective in the multiple people years being

---

*Deloitte refers to one or more of Deloitte Touche Tohmatsu Limited ("DTTL"), its global network of member firms, and their related entities. DTTL (also referred to as "Deloitte Global") and each of its member firms are legally separate and independent entities. DTTL does not provide services to clients. Please see www.deloitte.com/about to learn more.

saved in automation for testing and operations, but also for Deloitte client services, such as bringing in data from multiple sources and driving improvements in speed and quality," says Quintin.

Speaking the Language of Business

It's often been said that successful CIOs must speak the language of the business. A critical component in doing so requires technology executives to be able to interpret and clearly communicate to the CEO and fellow members of the C-suite how technology can be leveraged to help drive business transformation and create new revenue-generating opportunities.

"The modern CIO should be able to translate technology innovation into business opportunities that can be monetized as part of the transformation of commercial products or services," said Dave Roberts, CIO at Radius Payment Solutions in Crewe, Cheshire, U.K.

I spoke recently with Dave and Jon Wrennall, Group CTO at Advanced, the third largest British software and services provider, to gather their thoughts on the role of the CIO in driving business transformation and cultivating customer centricity. Here's a lightly edited transcript of our conversation:

Hunter Muller: What should be the CIO's role in helping to identify and then execute on business transformation opportunities with fellow members of the C-suite?

Jon Wrennall: Today's CIO should play a leading role in digital transformation, which I believe is at the heart of successful business transformation. The rise in digital disruptive technologies, like cloud computing, artificial intelligence, and automation, is already transforming and optimizing existing business processes and even enabling completely new business models. It's this innovation that every business large and small must invest in—and these innovations give CIOs the capabilities to help organizations reinvent themselves like never before.

The first opportunity is undoubtedly to drive productivity, which will free up the workforce to focus on value-added activities, eradicating the mundane for humans. And with the right digital transformation—driven by the CIO—a connected infrastructure will provide real-time business insights, which will also help to recharge the role of leaders. So, by starting with digital transformation, in turn the fellow members of the C-suite will work hand-in-hand to review roles and processes—essentially, looking to reshape their organization and making it fit for the future.

However, today's CIO must also recognize that digital is pervasive, with every business increasingly becoming a software business (or at least powered by one). Stakeholder management is therefore more complicated, so CIOs are under pressure to drive real business engagement and value, and ensure fellow members of the C-suite get on board to understand the opportunity and embrace it—or risk being left behind.

Dave Roberts: The CIO provides the technical direction and strategy that underpins and supports the growth strategy of the organization, providing insight and thought leadership with fellow C-suite members. It is the responsibility of the CIO function to investigate and research the technologies that are relevant to the organization's core market and that help fuel business growth and transformation.

CIOs that can successfully demonstrate these skills are perfectly placed to move into COO, nonexecutive director, or even CEO roles going forward.

HM: Why is it so important for CIOs to be more proactive than they historically have been in communicating opportunities for improving the business (driving efficiencies, creating new business models)?

DR: The CIO role is often varied with a wide variety of expectations from service delivery excellence, to driving better process efficiency, to then identifying how emerging technologies can differentiate and drive new business models and services. CIOs need to ensure they are demonstrating and communicating the value of the projects and products developed and delivered by the IT function.

Communicating this value helps move IT up the value chain from being an overhead function to a revenue- and IP-generating business unit. Proving technologies within an MVP model is often the most efficient way to demonstrate the value to the board with a "fail fast" approach to development. Innovation that can demonstrate a credible ROI and value to the business can then be scaled accordingly to either drive greater efficiencies or create new business models of operation.

JW: Critical to any digital transformation is people. People innovate, and, without them, businesses won't be able to innovate! So perhaps CIOs have been guilty of not effectively communicating the opportunities for improving in the past, but I'd argue it is probably more about being listened to and ensuring people understand the value.

In the meantime, two key realities have happened: Millennials have entered our workforce and demanded—expected—a more digital culture with the right tools to do their jobs. And the narrative about "digital transformation" has become more pervasive. The C-suite, who make the decisions about investment in technology, increasingly understand the critical nature of driving efficiencies and creating new business models, because they see new disruptive digital-first competitors entering the market and stealing market share. The role of the IT department is recognized increasingly as a business enabler.

I believe that what is holding this mass adoption back is that the latest digital innovations are often seen as the domain for large enterprises. This isn't the case and, while some SMEs are breaking boundaries, the reality is that most small businesses are still behind in digital transformation.

If the government along with leading technology voices and bodies can work harder to support small businesses—which make up 99.3 percent of the UK's private business sector—we'll see every size and shape business take advantage of artificial intelligence, machine learning, natural language processing, robotic process automation, and predictive analytics.[3]

Five years from now, every business will be digital in some form. Technology will be recognized as a competitive advantage as well as the enabler for growth and change.

HM: What are some ways in which the CIO can help foster a culture of customer centricity, not only within the IT organization but across the enterprise?

JW: Digital is disrupting how businesses need to deliver effective customer service. Engagement with customers is transforming as they become more digitally savvy. Just as organizations want a real-time dashboard to glance at vital statistics across their business, customers are looking for the same level of information on the channels they wish to operate. CIOs can act as the catalyst for change here to ask the right questions and demonstrate how technology can drive insight to help foster the culture of customer centricity.

For example, most businesses are grappling with how to keep ahead of their customers' needs, to maintain and delight them in every interaction. So, providing a digital face to a business is fast becoming the norm. In the same way as "born-digital" Millennials have an expectation about using digital devices to empower them in the workplace, canny customers are now demanding the same level of service across all relevant digital touchpoints.

Increasingly, businesses failing to embrace these changes are missing out as customers turn their back on brand loyalty and instead look for service that is personalized, value-added, flexible, and demonstrates innovation. The results of the Advanced "Annual Trends Survey Report 2017/18" provided insight into the increasing use of social channels:

- 49 percent say social media has enabled them to improve or innovate the way they interact with customers, up from 45 percent in 2016.

- 48 percent now use social media to learn more about new services and suppliers.

- At 74 percent, LinkedIn tops the social channels for learning more about new services or suppliers.[4]

These digital touchpoints are critical for driving customer service and proactive engagement. The maturing of technology to do things we could only dream of a few years ago means that, with increasingly accessible tools, people can drastically reduce the time to deliver the results of yesterday's experts. Core FMS and ERP systems—which have been used effectively to manage and run a business—are now transforming in the cloud to move beyond traditional historic and forecasting reporting, to provide business intelligence and analytics. With the introduction of deep learning, the inclusion of augmented intelligence, business process management, and robotic process automation,

CIOs can help the business start to unlock this data, enriching and creating new insight—enabling new customer-centric business models to be created.

The question I'd encourage CIOs to ask fellow members of the C-suite is this: What is the potential cost to the customers of your business if you do not have a digital-first strategy and/or vision?

DR: The culture of customer centricity means putting the customer at the center of everything you do and challenging the approaches taken to ensure they align back with the overall strategy.

We look at the overall experience of our products and services and how we service each touchpoint on that journey and the methods of interaction we provide to customers. Our products and services are developed around what the customer needs to service their own businesses and allowing them to focus more on their core activities in an efficient manner. The aim should always be to enhance the customer journey and value through either the efficiency or experience of using your product or service.

The CIO function can help to supply the tools that provide better customer insight through the use of CRM, business intelligence, AI, and social platforms, ensuring that the customer experience is relevant, efficient, useful, and accurate. Developing products and services with the customer in mind from the onset will ensure that the experience is engaging, intuitive, and the value proposition is clearly recognized.

Organizations that take advantage of technology are able to feel more connected with their customers with a greater level of authenticity to drive a culture of customer centricity.

Notes

1. Geoff Webb, "The Evolving Role of the CIO in 2018," January 9, 2018, https://www.forbes.com/sites/forbestechcouncil/2018/01/09/the-evolving-role-of-the-cio-in-2018/#3cc168fc1c8e

2. "Role of CIO Is Changing and Growing in Importance, Say New Forbes Insights Studies," Press Release, March 28, 2018, https://www.forbes.com/sites/forbespr/2018/03/28/role-of-cio-is-changing-and-growing-in-importance-say-new-forbes-insights-studies/#27164441426c

3. 2020 National Federation of Self Employed & Small Businesses Limited, "UK Small Business Statistics," 2020, https://www.fsb.org.uk/uk-small-business-statistics.html

4. Advanced, "Advanced Annual Trends Survey, 2017/18," https://www.oneadvanced.com/trends-report/2017-18/

Chapter 4

Macro Challenges for Tomorrow's Executive Leaders

There seem to be plenty of jobs around, but somehow it feels as if something isn't quite right. From our perspective here in the technology sector, we have to ask ourselves if the jobs numbers are obscuring deeper challenges. "The country faces a tight labor market, with a million more job openings than there are eligible workers to fill them. That dynamic is complicated by a pool of workers that lacks the skills for some of the jobs available," wrote Jeff Cox of CNBC.[1]

As Cox notes, there are plenty of unemployed job seekers "who have skills but can't seem to find the right jobs to fit their qualifications."

The mismatch between openings and applicants is especially problematic in the tech sector. Ginni Rometty, who served for more than eight years as the CEO of IBM, recently spoke to Lori Ioannou of CNBC about the importance of using artificial intelligence to match applicants with jobs they will be passionate about doing. She also said she expects "AI to change 100 percent of jobs within the next five to 10 years."[2]

Rometty's comments highlight a major trend that all of us in the tech industry should be watching closely. I believe that, as technology leaders, we have a genuine responsibility to stay ahead of the curve.

"Rometty's call to action comes at a time when the AI skills gap and the future of work exhibit a growing sense of urgency," Ioannou wrote. "The technology sector accounts for 10 percent of U.S. GDP and is the fastest part of the American economy but there are not enough skilled workers to fill the 500,000 open high-tech jobs in the United States, according to the Consumer Technology Association's Future of Work survey. Yet the tech industry is concerned that school systems and universities have not moved fast enough to adjust their curriculum to delve more into data science and machine learning. As a result, companies will struggle to fill jobs in software development, data analytics, and engineering."[3]

The continuing success of our industry—success that is absolutely critical to that of the larger economy—depends on our ability to recruit, hire, and retain the best workers. In the technology space, acquiring top talent is the ultimate competitive advantage. We need the smartest minds and the most innovative thinkers to drive our industry forward.

Are we building a talent pipeline? Are we allocating the proper resources for education and training? Are we adopting new models for developing new skills and abilities? Are we providing clear leadership and a strong voice so our elected officials truly understand the needs of the tech economy?

Technology has made our lives easier and more prosperous. But we need to start looking farther down the road and making absolutely sure we are building the foundations for continuing growth and success. As technology leaders, we have a responsibility to the future. Let's fulfill that responsibility.

Creating Wealth and Driving the Global Economy

Despite ongoing gyrations in the stock market, faith in tech companies has remained strong. What's generating those high levels of investor confidence?

I believe the answer is fairly clear: Over the past thirty years, tech companies have delivered unparalleled value to investors, consumers, and the broader economy. In today's turbulent markets, tech stocks are considered safe bets. That would have been unimaginable ten years ago.

People generally feel good about technology. They trust technology, and their trust translates into phenomenal revenue for the tech companies. The love affair with tech extends far beyond mobile phones. Sure, we love our phones, but we also love the tech in our cars and in our homes. We love that our health-care providers increasingly use technology to cure our illnesses and take care of our bodies.

That's the main reason tech stocks have become safe calls. As a society, our hunger for new and improved technology seems insatiable. Hunger drives demand, which in turn drives sales and profits. That's why markets love tech stocks.

What could possibly go wrong? I can think of a couple of ways in which investors might grow wary of tech stocks. A series of major cyberattacks could rattle our sense of confidence in tech. A couple of fatal accidents involving self-driving cars might cause people to think twice about the risks of too much automation. Widespread power outages or water shortages would diminish our trust in technology.

Another potential danger is the unchecked use of proprietary algorithms in a growing number of services used by large organizations such as banks, insurance companies, credit bureaus, and government agencies. As predictive analytics and artificial intelligence are baked more deeply into IT systems everywhere, the level of risk posed by broken, outdated, or biased algorithms rises dramatically.

As technology leaders, we have a moral and ethical responsibility to point out the weaknesses in our systems. We need to make sure that our systems are robust, resilient, flexible, and adaptable. We also need to make sure the information we deliver is accurate, unbiased, and fair. We should advocate for solutions that are transparent and explainable.

We cannot allow technology to become a "single point of failure" in our rapidly expanding global economy. As the markets have clearly demonstrated, technology is universally perceived as a driver of growth and income. To an astonishing degree, technology has become the foundation of modern life. We simply cannot live without it.

Stock markets are more than just places for trading shares in companies. They are proxies for our collective vision of the future. Today the markets are telling us that technology is absolutely critical. I completely agree, and I urge you to consider the profound implications of the incredibly interlinked relationships between technology and the global economy.

Great Leadership Is Critical to Success

During World War II, the United States created the Manhattan Project, a top-secret effort to develop the first atomic bomb. The legendary project had two leaders: J. Robert Oppenheimer and General Leslie Groves. Oppenheimer was scientific, philosophical, and imaginative. Groves was rough, abusive, and demanding. Oppenheimer provided creativity and insight, while Groves provided focus and energy. Together, they succeeded and invented the age of nuclear warfare.

All companies and organizations can learn a lesson from the Manhattan Project. In most cases, you need more than great technology and superior processes to succeed. You need great leaders who are brave, creative, and audacious. Does your enterprise have leaders who are ready to reimagine and reinvent the business to drive growth and create value? In today's markets, the genuinely great companies will be the ones with the best and most courageous leaders.

Why U.S. Executives Should Pay Closer Attention to the Surging Chinese Tech Sector

China's economic power is growing, but the growth is neither monolithic nor guaranteed. Under the surface, there are choppy crosscurrents, dangerous riptides, and strong tidal flows that could overturn our expectations.

For companies in the United States and Europe, the rise of Chinese technology is a double-edged blade: The modernization of China creates more business opportunities for tech firms based in the United States and Europe. It also creates more competition for those firms.

Many observers have predicted that China will soon become the dominant player in global technology. China has clearly expressed its desire to become the world leader in tech, but the truth is that the country is still playing catchup.

Despite the rise of tech giants such as Alibaba, Tencent, Didi Chuxing, Ant Financial, and Huawei, China does not pose a serious competitive threat to legacy providers of enterprise infrastructure and services. That could change, but it seems unlikely that established titans like Microsoft, Google, Oracle, IBM, and SAP will simply roll over and allow China to gobble up chunks of their traditional markets.

It's also important to remember that the competition extends beyond a simple East vs. West scenario. India and Indonesia are home to a new generation of innovative tech firms that are poised to move swiftly into emerging Asian markets.

China's meteoric rise is by no means a guarantee of steady or consistent growth over the long term. The country's success is truly amazing, but there are hazards in its future. A new whitepaper from the Organisation for Economic Co-operation and Development predicts that China's demographic policies will result in a sharp economic decline soon after it surpasses the U.S. economy in 2030. After that, the United States will return to the top spot in the global economy.

"The forecast for the U.S. to outstrip China is not a prediction of any economic miracle in America—just an acknowledgment that China has set itself up for a brutal demographic collapse," writes Daniel Moss of Bloomberg. "Shortly after China overtakes the U.S. economy in size, all the legacies of the one-child policy coalesce as the society seriously ages, stalling out the Middle Kingdom's expansion. The United States will face demographic challenges, too, but nothing like this."[4]

It's important for technology leaders and executives to understand the nuances and complexities of our relationship with China. It would be foolhardy to ignore China or to underestimate its potential for disrupting traditional markets.

At the same time, it would be imprudent to ascribe too much power to China. For the time being, the best course is keeping a close eye on the nation's rapidly growing tech sector and scanning the horizon for competitive opportunities as they arise.

Top Tech Trends Include Increased Reliance on Social Networks

We live in an exciting period of history, and sometimes it's easy to get caught up in the latest fad. That's a natural human reaction, but it's rarely helpful when developing long-term strategy.

Now is the perfect time to take a deep breath and look ahead. Based on our research here at HMG Strategy, these are the top tech trends to watch:

- Technology will continue driving the economy and radically transforming markets worldwide. The digital revolution is still in its early stages, and there's a long runway ahead of us.

- The sheer strength and economic force of technology will translate into another great year for tech companies and tech stocks. From my perspective, now is a good time to buy tech stocks. The market is soft, price-to-earnings ratios are low, and many tech stocks are underpriced.

- State-sponsored cyberwarfare will escalate. As a result, cybersecurity will become a major priority for corporate boards and directors. To put it bluntly, however, most boards do not truly understand the risks of operating in a digital world. It's no longer enough to have a strong perimeter; the bad guys are already inside your organization. The cyberdefense strategies of tomorrow will include "proactive defense," which involves collecting intelligence about your adversaries and genuinely understanding their strengths and weaknesses.

- Artificial intelligence will penetrate more deeply across the modern enterprise, and users at all levels will expect to see AI capabilities baked into more products and services. That said, tech executives will continue struggling with AI's value proposition. Some companies will integrate AI successfully into their business models and achieve new competitive advantages. Many companies will feel obligated to invest in AI, though, even if they're not sure how to use it.

- The transition from traditional data centers to the cloud will continue and accelerate. The caveat here is that many corporate technology executives do not yet fully grasp the basics of "cloud economics," and this lack of understanding may lead to some embarrassing mistakes. The cloud is unquestionably the right place for most data, but tech execs need to ask tough questions and get ahead of the knowledge curve so they aren't surprised when the bills come due.

- More and more transactions will shift to mobile. We're already seeing "mobile-first" e-commerce organizations, and soon we'll see "mobile-only" companies in the market. The shift toward mobile will accelerate, completely transforming the face of e-commerce.

- Organizations in every sector of the economy and at every level will move the bulk of their communications to social channels. Within the next couple of years, social media will become the primary source of information for most people. The implications of this trend will be widespread and inescapable.

- Customer focus will become the primary strategy of companies and organizations in every industry and in every market. The idea of product-centricity will continue to fade as the customer-centric service economy grows in scope and scale.

Each of these predictions involves massive amounts of technology and technology-based services. We live in a world of connected digital technologies. The tech industry will continue to thrive, and the average value of tech stocks will reach new heights in the years ahead.

Making the Future Global Energy Company

Disruption is affecting companies in nearly every corner of the economy, and the energy industry is no exception.

Meanwhile, technological innovation is contributing to the falling costs of renewable energy and energy storage.

As VP and CIO at Shell Downstream, Craig Walker and his team are immersed in these disruptive changes affecting the energy industry. I spoke recently with Craig, a chairperson and a speaker for at HMG Strategy's 2019 London CIO Executive Leadership Summit, to get his perspective on the CIO's role in helping to guide disruptive change within the enterprise. Here is a lightly edited transcript of our conversation:

Hunter Muller: What should be the role of the CIO in working with the CEO and the board to guide the company toward its future state objectives?

Craig Walker: You can no longer be the head of technology and waiting to provide service when asked. Every single person in the IT organization needs to understand how the company makes money. You need people in IT to have a commercial mind-set. You are a businessperson first.

Finding people with that mind-set is not easy. The other thing IT people find hard to do is to say no. You need to have the credibility to be able to walk into a business meeting and say, "Sorry, this is not the way we should do this," and then explain why or offer alternative options.

HM: How do you go about building these business skills among your IT team?

CW: We have people who are called IT managers but they're also known as the business interface. They don't have day-to-day responsibility for delivery of IT services, but they do have accountability to the business for all IT does and to be highly visible in the business and play a lead role in shaping strategy and making tactical decisions in business meetings.

We've held workshops around the world to develop the business skills that our IT team members need. We've taught them a lot about stakeholder management, value delivery, and innovation. They've been on quite a journey. This has involved training, with plenty of roleplay, to help guide a conversation with a senior executive, how to take control of a situation, influence, explain, etc.

It's been a difficult journey for some, and others have thrived in it. Some have been moved back into more technical roles because they're more comfortable there.

I want people to spend 10 percent of their time learning. You have to constantly learn in today's environment. And it's tough because the pace of change is occurring so quickly.

Our whole premise is changing. We're moving from being a hydrocarbon company to an electron company. And IT is a big part of it. We won't thrive in this energy transition if the IT does not deliver.

HM: How do you reimagine and reinvent? Where do you get the inspiration?

CW: I think I was lucky in my career in Shell that you get sent to offices overseas and you're given the autonomy to get things done. I've always been turned on by the business. I like the technology, but it's all about the business outcomes.

Each of the people who work for me would be a Top 50 CIO for another company. I have a fantastic team under me. I've had a chance to step back and think about how to my team and technology can help Shell thrive in the energy transition. And everything involves IT. For over a hundred years Shell has been at the cutting edge of innovation, and this has allowed us to move with the times. Now, as we are working on the challenge of bringing more and cleaner energy to the world, we are investing in the new energy solutions and have set the ambition to cut the net carbon footprint of our energy products by around half by 2050. And we fully support the Paris Agreement. Playing a key role in delivering a sustainable energy future is a big inspiration.

HM: Can you point to an example of what Shell is doing over and above its traditional oil and gas business?

CW: It is about moving from selling great products, to delivering great services wrapped around those products. Through Shell Energy we provide nearly one million homes in Great Britain with renewable energy.

Shell's NewMotion is one of Europe's largest electric vehicle charging providers to homes and businesses.

In early 2019, Shell announced the acquisition of Greenlots, a California-based company that provides electric vehicle charging points, charging network software, and grid services across the United States. Greenlots also has a growing business in Canada, Thailand, Malaysia, and Singapore. With NewMotion and Greenlots, Shell is developing flexible home and workplace charging solutions, including for fleets.

At the back end of this you have our traditional retail sites, which also increasingly provide electric charging through Shell Recharge. What I really want to do is if you're a homeowner and you show up at a Shell charging station, I want to be able to recognize you immediately. We want to be able to provide customers with differentiated services they've never seen before. Take care of their energy needs whether at home, work, or on the move, and make the experience as seamless as possible.

HM: Which technologies hold the most promise for you at Shell?

CW: What we're doing at the moment with machine learning and AI has great possibilities. Where we have big data sets, AI is performing for us. Where we need to do more with AI is in the use of our customer data. With AI, you've got to pick the right places to use it and to use it carefully. The goal is to apply AI against customer data to identify opportunities for providing a frictionless customer experience.

We've got to become a customer-centric organization. What works for customers in Malaysia may not work for them in South Africa or New York. It's easy to say but ridiculously difficult to do.

I'm working very nicely with our downstream directors who are leading the customer-centricity drive. You really need to step back and listen to what others are saying. We have always been known for our high-quality products; now more than ever, our customers' needs and expectations are at the heart of everything we do, and that is not an easy journey.

We must demonstrate, through actions, that Shell is an energy company that is striving to deliver cleaner energy solutions to the world; and all of that will be enabled by excellent and affordable IT solutions.

Fine-Tuning the Systems Experience to Deliver an Optimal Customer Experience

As many customer practitioners say, a smooth customer experience begins with a solid employee experience. After all, if a customer-facing employee has trouble finding information for a customer, the employee will likely not deliver an exceptional customer experience.

This helps explain why 53 percent of customer experience professionals surveyed by West Monroe Associates found that a motivated and equipped workforce are critical to achieving an improved customer experience.[5]

With the goal in mind of probing this fascinating insight, I spoke my good friend Ashwin Ballal, PhD, SVP, and CIO at Medallia, and asked him to share his perspective on the important connection between employee experience and customer experience. The following is a lightly edited version of our dialogue:

Hunter Muller: Please explain your focus on the "systems experience" and what it entails.

Ashwin Ballal: Our lives are defined by our overall experiences, both at work and home, as consumers. Therefore, CIOs and IT organizations that recognize these experiences and support their employees in these moments that really matter will get to acquire and retain top talent, increase productivity and collaboration, and be innovative so as to win in the future. My fundamental belief is that with any enterprise, small or large, employees are subjected to a large number of systems experiences that are suboptimal. Most of the systems have been supported and managed by the IT organization. These systems experiences have a large impact on the employee experience, which in turn has an impact on the customer experience. This is true whether it involves communication, collaboration, or any other business automation systems.

HM: How is this approach beneficial in the modern enterprise?

AB: How you capture feedback from the different stakeholders in moments that matter can and will give the

CIO and the IT organization rich insights and empathy for the systems experience that they're providing to their users. CIOs and their teams need to determine which systems experiences are super-critical and know that they can capture real-time feedback at any time and close the loop so that these systems experiences can be great for employees.

HM: What are some of the benefits that you've achieved at Medallia from a systems experience mentality?

AB: I've been able to take the Medallia platform that captures the essence of each one of my constituents' systems experience feedback and in turn use that feedback to make the necessary strategic changes that might be needed to provide great system experiences. Since we pulse our employees for feedback, it is the Net Promoter Scores that matter on how IT is improving the systems experience over time. We have seen a three times increase in response rates for our request for feedback, an increased visibility within my organization to the pulse of the employees, and an empowered IT team that can take action in the moment that matters to employees.

HM: What are some recommendations for improving the employee experience?

AB: You have to walk in the shoes of your business user. Do you have empathy for that businessperson? Take a salesperson, for example. What systems do they use and how do they access those systems? Do they have to enter data manually? Can that experience be modified or automated to make it better and a delightful experience for them?

Living a day in the life of our users and understanding how they use a particular technology can help us to enable them to become more competent in their roles.

HM: What are some initial steps that fellow CIOs and other executives can take to shape a systems experience program?

AB: The first step is measuring baseline performance. In many systems experience programs, we want to get the feedback on the top-end systems that most employees engage with and get their feedback on those experiences.

I'd also recommend putting together a focus list of critical systems to get started and ask about the systems experience of the top-ten core systems that employees use and how they feel about those systems and then propagate to other systems after that. Establishing a baseline performance before digging deeper into systems experience is critical.

HM: What advice can you offer for empowering the organization to deliver improved customer experiences?

AB: We've historically been focused on support instead of the overall experience. Try it yourself before you subject your users to the same experience. Changing that reactive mind-set to a more proactive mind-set can lead to great employee experiences.

Notes

1. Jeff Cox, "America's Bustling Jobs Market Is Still Leaving Some People Behind," CNBC, April 5, 2019, https://www.cnbc.com/2019/04/05/americas-bustling-jobs-market-is-still-leaving-some-people-behind.html

2. Quoted in Lori Ioannou, "IBM CEO Ginni Rometty: AI Will Change 100 Percent of Jobs over the Next Decade," CNBC, April 2, 2019, https://www.cnbc.com/2019/04/02/ibm-ceo-ginni-romettys-solution-to-closing-the-skills-gap-in-america.html

3. Ioannou, "IBM CEO Ginni Rometty."

4. Daniel Moss, "OK, So China Will Surpass the U.S. Economy. Then What?," Bloomberg Opinion, September 2, 2018, https://www.bloomberg.com/opinion/articles/2018-09-02/what-happens-after-china-surpasses-the-u-s-economy

5. Quoted in Phil Britt, "The Intersection of Employee Experience and Customer Experience," October 29, 2018, https://www.cmswire.com/customer-experience/the-intersection-of-employee-experience-and-customer-experience

Chapter 5

Confronting Global Shifts

Technology has become absolutely indispensable to virtually all of us, everywhere in the world. We all depend on technology. It's become our ubiquitous 24/7 servant, guide, and resource.

We use technology constantly. We interact with our tech every minute of every day. Many people even remain connected to tech after their day is over by using apps that monitor the duration and quality of their sleep.

Technology began as a support tool for business. Over the past seven decades, however, tech has evolved into a giant economic sector. By several estimates, tech is the now the third-largest chunk of the world's economy, generating nearly $7 trillion in spending.

Clearly, the tech industry is well positioned for continuing success over the long haul. From my perspective, that's certainly good news. Knowing that tech will ultimately prevail keeps me from worrying about day-to-day or even week-to-week fluctuations in the stock markets.

In a recent edition of the HMG *Tech News Digest*, I noted that chip stocks were taking a beating.[1] The companies mentioned in the item were AMD, Micron, Texas Instruments, and NVIDIA. Yes, those companies are running into some speed bumps. But frankly, I don't see a cause for concern. The need for chips and microprocessors will continue growing as billions more devices and millions of systems become part of the Internet of Things.

The rapid growth of the IoT alone will drive tech spending in practically every part of the economy and at every level. Five years from now, it will be hard to find a device or appliance that is *not* connected to the internet. Everything will be "smart," and everything will be part of a larger network. Connectivity is the future of our civilization.

The exponential rise in connectivity will create surges of spending across every industry. When you consider the possibilities, it's hard to imagine a scenario in which tech doesn't emerge victorious.

I'm not saying that everything will be rosy. There will be rough patches, for sure. Markets will experience bad days and bad weeks. But the technology sector will remain strong and will continue to drive growth and spending across the board.

CISOs Are Essential for Achieving the Future State

Historically, chief information security officers (CISOs) have been synonymous with helping to protect the enterprise and its information assets and helping to mitigate organizational risk. But with CEOs and board members primarily focused on growing the business, CISOs are increasingly being looked upon to help strike a balance between growing the business and pushing digital strategies forward while protecting the organization.

According to a 2018 Accenture study, CISOs and security leaders typically are not involved when business units develop new products, services, and processes—each of which entails some cyberrisk. On the bright side, 38 percent of respondents to the Accenture study say the CISO is brought in before a new business is considered.

For his part, Steven Booth, vice president and chief security officer at FireEye, has had a few experiences in guiding business enablement as a security leader. "One of our business units came to me and they had just gotten to proof of concept with a new service but they hadn't started writing code yet. They wanted to make sure they could do this project without any GDPR [general data protection regulation] or privacy issues. That's a great conversation to have as a CISO—that's the enablement and innovation side," says Steven.

"The CISO needs to understand the business from A to Z," explains Jason Hengels, founder of Exposure Security. "It's not a role that lends itself well to someone who just wants a job in the security space. To truly provide optimum value, the CISO needs to have a mind for business and must be able to understand the potential positive and negative impacts to the business that can occur due to their decisions."

As a simple example, anyone can say, "If we have a 7-character minimum password policy on our website, we might get hacked and that's bad," Jason says. "A good CISO needs to understand the potential revenue impact posed by making the password policy longer and the likely earnings impact if the business suffers a breach due to that particular issue. A CISO who can execute in that manner will make better decisions, and their recommendations are more likely to be followed by management."

Plus, as companies are increasingly "going digital" and executives explore what this means for their business models and approaches to customer experience, planning for and embedding the right level of cybersecurity into digital transformation initiatives becomes critical, says Chad Kalmes, vice president of Technical Operations at PagerDuty.

Chad, who is an advisory board member for the HMG Silicon Valley CISO Summit, points to how cybersecurity has historically been approached as a bolt-on to preexisting business processes. "A lot of that 'after-the-fact' approach honestly led to a great deal of user friction with initial security controls and projects. Security programs and controls weren't often planned into the strategy from Day 1, so they generally wound up being less successful and more burdensome on the end user. With so many companies now rethinking some of their core processes and approaches to how they reinvent and reimplement their strategies, and with growing trends like DevSecOps, CISOs have the opportunity to evaluate risks in conjunction with those changes and design the right kinds of preventative and detective controls along the way. That should lead to not only better risk reduction for the companies involved, but also a better and less obtrusive fit for end users and customers."

In order for CISOs to make this pivot, they also must break free of the "Doctor No" role.

"While the CIO role has matured, the CISO role is still relatively new, and it needs to evolve from a technical role to more of a business role," says Mark Egan, partner, StrataFusion. "Instead of being the person who says 'You can't do this,' the CISO needs to find ways for security to become a competitive advantage for their companies."

Among the priorities competing for the attention of CIOs in 2018, improving cybersecurity has moved farthest up the radar, jumping 23 percent from 2017 to 40 to 49 percent in 2018 in the Harvey Nash/KPMG CIO Survey 2018.[2]

A Truly Holistic View of the Enterprise

Thanks to their unique view across the enterprise, CIOs play a critical role in helping to reimagine and reinvent the business. These capabilities haven't gone unnoticed by boards of directors; in fact, 70 percent of boards expect their CIO to be an innovative force and a creative disruptor, according to a study by BT Global Services and Vanson Bourne.[3]

"CIOs have a great vantage point from which they're able to see all of the business and offer suggestions to fellow business leaders," says Renee Arrington, president and COO of Pearson Partners International, Inc. "The question is whether they have the strategic skill sets to identify and then clearly articulate those opportunities."

CIOs come to the table with a wide range of competencies and characteristics that can help their companies drive business transformation and find new opportunities for growth.

According to Hugo Fueglein, managing director of Diversified Search, "The CIO needs to break down the silos and think across the company in terms of functions, markets, and customers. It's a horizontal thinking process versus a vertical thinking process. Even in the IT organization itself in terms of using AI and machine learning for digital transformation, CIOs have to be thinking about how to access information both inside and out in a very holistic way."

Identifying and communicating opportunities for transforming the business also requires CIOs who are comfortable working with the executive team in formulating strategy. "Every IT executive search I've been involved with over the past five years has required a candidate who can have a seat at the table and work with the leadership team to apply technology to the strategy of the business," says John Fidler, managing director, Retained Search, Fidato Partners.

Candida Seasock, founder and president of CTS Associates, LLC, which serves as strategic advisors to CEOs and growth companies, believes that one of the ways long-standing CIOs can demonstrate the value they're delivering to the business is by creating a marketing plan that validates the business services the IT organization is delivering and the results that are being achieved.

"The CIO of today and tomorrow should be focused on meeting both internal business needs and external customer needs," Candida explains.

Some industry insiders believe CIOs need to become more proactive in communicating business opportunities to the CEO and the executive team than they had been in the past.

"That's one of the attributes of a very good contemporary CIO," Hugo says. "When we look for CIOs, we look for someone who is passionate about articulating effectively across any audience. This is where the use of data and analytics has to be orchestrated well to identify areas of opportunity."

"Who the CIO shares these opportunities with also depends upon the structure of the company," addsed John. "Whether it be to a member of the corporate executive team or a leader within a business unit, the CIO needs to be comfortable communicating opportunities as they present themselves."

Which executive the CIO communicates business opportunities to also depends on the personalities and receptivity of the individuals involved, notes Renee.

"Both of these factors play a role in communicating ideas and when to communicate," she explains. "You can share a ton of ideas, but the person who's receiving that information has to be prepared to hear it. You've got to recognize when the timing is right."

Of course, reporting relationships are also key in making an impact. According to Candida, "The CIO should be reporting to the CEO or the COO, depending on the size of the organization. If they report to the CFO, the CIO likely won't have a voice at the table in terms of supporting the business. If they report to the CFO, it becomes a cost conversation, not a value conversation."

How Artificial Intelligence Is Transforming IT

Elon Musk has made more than his fair share of news, mostly talking about Tesla and SpaceX. But I think we also need to focus on another aspect of Musk, namely his stated concerns about artificial intelligence.

Musk isn't alone in expressing anxiety about AI. The late Stephen Hawking also warned that AI could wreak havoc on a world that's not prepared to manage its potentially destructive power.

AI isn't just another "feature" that will be bolted onto existing systems. AI will quickly replace existing systems, rendering them obsolete and irrelevant. Once AI spreads through the IT universe, all of our roles as technology leaders will fundamentally change. In short, we will experience the disruption that we often see happening in other industries.

Imagine a world with no systems administrators, software developers, or business analysts. That scenario will become a reality sooner than we imagine. In the very near future, AI will be baked deeply into every conceivable system and platform.

How will IT leaders add value when most of IT becomes fully automated? That's a hard question, and we must begin considering it seriously. The most pressing issue is acquiring talent. You'll need a process for identifying, recruiting, hiring, and retaining people who understand AI and who know how to use it. You'll need to create appealing work environments to attract the best minds and keep them focused.

It's not too soon to begin setting up talent pipelines. Are you reaching out to local colleges and universities? Are you actively recruiting people with data science skills? If you aren't, you should be.

Mark van Rijmenam of the Netherlands wrote a good post recently on the difference between "good AI" and "bad AI."[4] In his post, he argues that when AI is applied thoughtfully and carefully, its benefits outweigh its potential for causing harm. But when AI is applied haphazardly or indiscriminately, it can morph into something genuinely dangerous.

"Good AI," he wrote, must be "explainable." In other words, it can't be a black box. The AI's decision-making processes must be visible and understandable to the human mind. In essence, we need to know how it works and how it's making decisions. When we don't require an AI's processes to be transparent and understandable, we're abdicating our responsibilities.

AIs feed on data, so we also need to make sure that our data sources are clean and unbiased. We've already seen instances in which biased data has led AIs to make biased decisions, so this isn't science fiction. It's already happening.

In aviation, pilots are taught to stay ahead of their airplane's power curve. Allowing a plane to get behind the power curve is an invitation to disaster since it won't have enough power to recover if a problem arises.

In a sense, we're allowing ourselves to fall behind the AI power curve. After a certain point, it will be impossible to recover if something goes wrong. That's not where we want to be.

Dealing Effectively with the Transfer of Sensitive Technologies

Most of us have no desire to see a full-blown trade war between the United States and China, so we tend to react favorably when the rhetoric is dialed back.

But a momentary reduction in tension won't resolve some of the most fundamental problems. A study by Michael Brown and Pavneet Singh highlighted in a recent edition of *The Information* raises serious questions about the extent of China's investments in emerging technologies.

What's even more alarming, however, is the lack of a coherent strategy here in the United States for dealing with China's ambitions for becoming a dominant force in the technology industry.

"The United States does not have a comprehensive policy or the tools to address this massive technology transfer to China," wrote Brown and Singh. The Committee on Foreign Investment in the United States "is one of the only tools in place today to govern foreign investments, but it was not designed to protect sensitive technologies and is only partially effective."

Moreover, wrote the authors, "the U.S. government does not have a holistic view of how fast this technology transfer is occurring, the level of Chinese investment in U.S. technology or what technologies we should be protecting."

In other words, we have an antiquated and inefficient system for monitoring foreign investments in sensitive technology. That is not good news for our industry. China is already working hard to dominate the market for rare minerals required for the batteries that power our smartphones and electric vehicles. Does it make sense to also sell China our most valuable intellectual property?

By 2030, China's economy will be 50 percent larger than the U.S. economy. That scale will create incredible leverage. And the Chinese are preparing for a future in which the United States is a subordinate player.

"China is investing in the critical future technologies that will be foundational for future innovations across technology both for commercial and military applications: artificial intelligence, robotics, autonomous vehicles, augmented and virtual reality, financial technology and gene editing," according to Brown and Singh.

I believe strongly in free trade and I welcome all forms of legitimate competition. But I genuinely believe we need to be aware of China's long-term goals and ambitions. Our lack of a holistic strategy for dealing with China poses risks of unimaginable magnitude. As an industry, we need to make our voices heard.

Tech Industry Will Remain a Strong Driver of Disruption

Unquestionably, volatility has become the new normal. We live in uncertain times, thanks to a continuing cycle of innovation and disruption. The economist Joseph Schumpeter called it "creative destruction," a term that perfectly captures the awesome energy and extraordinary force of modern markets.

The idea of continuous change frightens some people. But for visionary entrepreneurs and investors, the turbulence of our incredibly complex global economy is a gift that keeps giving. Every day presents new opportunities for those bold enough to seize the moment and ride the wave.

This unprecedented tide of transformational change is driven by the rapid development and application of new digital technologies. What's genuinely astonishing is that we're still in the early innings of a long game. The current era of technology-driven transformation will likely last another 15 years. There are indications that it might continue for several decades.

From a macroeconomic perspective, the transformation makes complete sense. A surprisingly large proportion of today's technology infrastructure is 30 to 40 years old. The old infrastructure is being replaced by platforms and applications that are newer, faster, more efficient, and far less costly. In addition to having smart cars and smart homes, we'll have smart power grids, smart cities, and smart nations.

That's great news for society and even better news for the technology industry. Don't listen to the naysayers. The tech industry is enjoying the most significant boom in its history.

From now until the mid-2030s, we will have our hands full with new projects. This is truly the best time to be a technology executive.

Digital transformation isn't a buzzword or a fad. It's an absolute imperative. Organizations that do not transform and disrupt themselves will perish. It's really that simple. Technology is the beating heart of disruption. Technology turns disruptive ideas into innovative products and services that conquer new markets.

I predict another fabulous year ahead for our industry. I foresee another year of amazing innovation, invention, disruption, and growth. Some segments of the economy may experience painful dislocations, but most of society will reap the benefits of a global economy that's more efficient, more powerful, and more equitable than ever before in history.

Validating the Belief in IT as a Value Driver

I always find it reassuring when a large and respected organization validates the core beliefs and central value proposition of our industry. A report by Capgemini showed that new applications of IT could add $512 billion in global revenue to the financial services sector by 2020.[5]

That's excellent news and timing is perfect, because it means there's still plenty of runway ahead for smart CIOs and visionary technology executives.

Specifically, the Capgemini report cited robotic process automation, artificial intelligence, and business process optimization as key technologies for driving unprecedented growth over the next four years. Those newer technologies are being used to boost revenues, create value, and open new markets for unparalleled growth.

"Leaders within the financial services industry have begun taking automation directly to their customers, using it as a revenue generator rather than just a cost saver," the report stated.

From my perspective, each sector of the economy has valuable lessons and experiences to share. As leaders, we're always looking for the next "new wave," and it certainly looks like the financial services industry has latched onto something important.

According to the Capgemini report:

[O]n average, over one-third (35%) of financial services firms have seen a 2–5% increase in topline growth from automation, with faster time-to-market and improved cross-selling efforts as the key factors that influence gains. Meanwhile, 64% of organizations from across different segments have seen improvement in customer satisfaction by more than 60% through intelligent automation.

The Capgemini report supports what I've been saying and writing for years: Visionary companies perceive IT as a value generator and business growth accelerator. They press hard for digital transformation and move ahead of the pack to deploy technologies that make a difference. They aren't afraid to take risks and push the envelope when they sense a competitive advantage.

I find it especially interesting that financial services firms are focusing their energy on pushing new technology directly to the consumer. The Capgemini report also found "that more than half of firms (55%) are focused on increasing customer satisfaction through automation, while close to half (45%) see growing revenue as a key objective."

The report definitely aligns closely with my central message of lead, reimagine, and reinvent the modern enterprise to drive growth and create value. The trends cited in the report offer amazing opportunities for courageous CIOs and technology leaders. Will you be ready to seize those opportunities when they arise?

Across the Atlantic: Content Protection Vote Fails, But the Battle for Control Isn't Over

We've been focusing on the trade war with China for quite a while, and now it's time to shift our attention briefly to Europe, where media companies have been lobbying strenuously for stiffer regulations over content. Even former Beatle Paul McCartney joined the battle, arguing in favor of stronger regulations to protect intellectual property.[6]

This issue is highly relevant to CIOs at companies all over the world. Sooner or later, lawmakers in the United States and Europe will adopt far more stringent laws protecting copyrighted materials that many people routinely post on their company websites. When those tougher rules go into effect, CIOs will become responsible for weeding out protected content and providing proof of compliance. The tougher regulations would have a broad impact, and IT departments likely will play a role in enforcing the stronger rules.

The continuing push–pull between the content providers and the tech companies returned to the spotlight recently, when the European Parliament voted by a slender margin to reject a proposed set of stronger rules governing the use of copyrighted content. The decision is a temporary victory for the tech firms and a setback for media companies that want more control over how their content is distributed.

The media companies contend that tech giants such as Facebook and Google have far too much leeway in their use of content from media producers. The media firms were pressing for regulations that would require websites to use special algorithms to detect and block unlicensed content. It would have also forced the websites to pay for articles, images, and music posted online.

"Media companies have been seeking a rewrite of Europe's copyright laws that would give them more power to restrict how their content is distributed. They also cited concerns that

Silicon Valley was not playing a strong enough gatekeeper role when it came to curtailing hate speech, violent extremism and fake news," wrote Adam Satariano of the *New York Times*.[7]

The European Parliament's vote followed on the heels of an unusually vigorous campaign by the tech companies to paint the enhanced regulations as dangerous to internet freedom. Evidently the European lawmakers were swayed by those arguments.

"After a well-coordinated effort by companies including Facebook, Google, Reddit and Wikipedia, as well as a grass-roots campaign by backers of an open internet, the European Parliament on Thursday [2018] rejected the proposed copyright law. Though lawmakers can still revise the bill and call another vote, the result is a blow to media companies that had believed that, if ever there was a good time to impose tougher rules on tech giants, this was it," Satariano wrote.

Facts and Fiction: Talking to the C-Suite and Board About AI

For a quite while now, I've been hearing people tell me that AI will soon replace most of the jobs in IT. Although there is a grain of truth in these prognostications, for the most part they are misleading and generally untrue.

As a senior technology leader, you undoubtedly will be asked if it's true that AI will soon eliminate the need for human workers in IT. Most experts seriously doubt that AI will lead to wholesale job losses in the near future. So why are people suddenly talking about 90 percent reductions in the IT workforce?

The claims made recently that AI would displace many IT workers is based on the mistaken belief that robotic process automation (RPA) is the same as AI, which it is not.

It's conceivably true that RPA might eventually replace large numbers of technology workers, but it's more likely that AI will create new jobs faster than they can be filled.

As its name suggests, RPA is about automating processes. AI, in contrast, is about sifting through data for insights. RPA is based on preconfigured rules that enable software robots to execute fairly straightforward and predefined processes. Modern AI is based on machine learning, a technique that essentially enables computers to teach themselves how to think. That's a significant difference.

Here's a great quote from the CFB Bots blog: "For simplicity, you can think of RPA as a software robot that mimics human actions, whereas AI is concerned with the simulation of human intelligence by machines."[8]

AI is almost the opposite of RPA. Instead of following a rote plan, AI generates value by discovering invisible patterns in huge mountains of data. That's why AI probably will create more jobs than it eliminates.

In other words, AI creates value and RPA saves money. Both are good, but they're definitely not the same. In fact, they are two ends of a spectrum. RPA takes the human drudgery out of work, while AI discovers new kinds of work for human beings to perform.

How Does Facebook's Monopoly Impact Innovation in the Technology Space?

I remember learning about monopolies in high school. As I recall, the monopolies we studied were legendary companies, such as Standard Oil, American Tobacco, U.S. Steel, and the original AT&T.

Based on what we learned in school, the idea that a social media company would ever be considered to be a monopoly seems far-fetched. But according to a comprehensive article by Dina Srinivasan in the *Berkeley Business Law Journal*, Facebook has risen to a level of monopolistic power.[9]

Examining and debating Facebook's role in the modern global economy seems especially relevant at this moment. In March 2019, Facebook CEO Mark Zuckerberg posted a lengthy message in which he said the company would shift its longtime strategy and refocus its effort to develop more products and services that enable private, encrypted communications.

Zuckerberg's post, "A Privacy-Focused Vision for Social Networking," apparently signals a sea change in Facebook's long-term vision for the future.[10] But it doesn't clear up larger questions about the company's sense of responsibility to its users, many of whom are genuinely worried about how the social network has compromised their personal privacy.

In her article, Srinivasan made a compelling case for the idea that Facebook indeed has become a monopoly and that its ascent was propelled by monopolistic business strategies.

Why should senior technology executives pay attention to Srinivasan's argument? I believe that she raises a legitimate concern that has an impact beyond social media. Here in the technology space, we are well aware of the risks that can arise when companies achieve monopolistic leverage in fundamental areas of tech.

It's not just a matter of large companies gaining incredible advantages in their markets. Monopolistic practices also hamper the entire tech community by reducing competition and stifling innovation.

Here's a good example: When Microsoft had monopolistic power, it was far less innovative and competitive than it is today. Microsoft's descent from its near-monopoly status has become a benefit for the market and for Microsoft.

Another example: The breakup of AT&T opened the door to decades of amazing innovation in the telecom industry, creating thousands of entirely new technologies and millions of new jobs all over the world.

When the U.S. government acts to break up monopolies, it's not acting on a whim. There are long-term consequences, both intended and unintended.

From Srinivasan's perspective, Facebook's rise to dominance was based on more than the innovative use of modern digital technologies. According to Srinivasan, Facebook has a track record of misleading consumers about its true intentions. Facebook started out by promising to serve as a trusted steward for personal data. But a series of unfortunate events occurring over the past couple of years have shown otherwise.

That said, people who like to use social media have limited choices outside of Facebook, which has become the default social network for billions of users.

If Facebook wants to be perceived by the market as a strong competitor but *not* as a monopoly, what should it do? Here is one of Srinivasan's recommendations: "Facebook should migrate from a closed to an open communications protocol. A user on Facebook should be able to send a message to, or receive a message from, a user of a competitive social network—in the same way that users of AT&T can call or text a user of Sprint, Verizon, or T-Mobile."

From my perspective, Srinivasan's observations and suggestions seem reasonable. At least they are a first step, and they might open the door to more competition and greater innovation across the tech space.

Are You Prepared for the Era of Content Validation?

Traditionally, CIOs have been responsible for maintaining and operating systems of records. Someone else—usually in finance or accounting—was responsible for ensuring the accuracy and veracity of the data in those systems.

We assumed the data was mostly accurate and that if we did a good job of managing the data, we could provide users with a "single version of the truth."

But that assumption—that the data is mostly accurate—has been undermined in recent years. The term "fake news" doesn't just apply to the news we see on cable television. Any data set can be altered or "faked," and that's a problem we all need to confront.

Today, most of us are keenly aware that much of what we see or read has been edited, altered, massaged, or spun to reflect a particular viewpoint. We no longer blindly trust the information we receive.

Since the dawn of science, data has been the gold standard of truth. Suddenly and unexpectedly, however, people are questioning the central primacy of data.

How does that shift in belief impact our role as stewards of data and information? It's too early to tell, but it seems that our responsibilities will have to evolve to keep pace with the changing perceptions about information.

I recommend reading a fascinating article in *Fast Company* by Hootsuite CEO Ryan Holmes.[11] In the article, Holmes argued forcefully in favor of taking stronger steps to ensure the accuracy and integrity of content. From my perspective, content includes data and information.

"The way forward isn't just an algorithm tweak or a new set of regulations. This challenge is far too complex for that. We're talking, at root, about faith in what we see and hear online, about trusting the raw data that informs the decisions of individuals, companies, and whole countries. The time for a Band-Aid fix has long passed," Holmes wrote.

Holmes foresaw the growth of a new industry based on content validation. If his prediction is accurate, CIOs and other senior technology executives will be responsible for evaluating, selecting, and deploying the most appropriate content validation solutions for their companies.

It's likely that many of the solutions for validating content will be based on artificial intelligence, which has a knack for spotting suspicious patterns in data. But reliable AI solutions require machine learning processes, which feed hungrily on mountains of data. How do we verify and validate the data used in the machine learning processes?

We all remember GIGO, which stands for "garbage in, garbage out." Somehow we let ourselves believe that we had solved the GIGO problem. Now it seems to have returned with a vengeance.

Taking Stock of Elon Musk

I highly recommend Charles Duhigg's long and excellent article about Elon Musk in *Wired*.[12] It's worth reading carefully, and I urge you to share it with your colleagues and peers.

A lot has been written about Musk and his efforts at Tesla to create a genuinely superior automobile. Duhigg, who was part of a team at the *New York Times* that won the 2013 Pulitzer Prize for explanatory reporting, described Musk as an incredibly complex individual with the strengths and flaws of a mythological hero.

In Greek tragedy, the hero pays a heavy price for his brief moments of triumph. It remains to be seen whether Musk achieves truly heroic status or lapses into irrelevance as he struggles to accomplish his self-appointed mission.

On some days, Musk seems like a man possessed by a noble vision to save the planet. On other days, he acts more like Don Quixote, tilting at windmills and raging at the world he wants to preserve.

The stakes are high. Musk isn't merely a popular icon; he's the CEO and cofounder of Tesla and SpaceX, two major companies that are striving to alter the course of humanity. If Tesla fulfills its long-term potential, our dependence on fossil fuels may be greatly reduced. If SpaceX succeeds, our grandchildren might be living on Mars. Those are not trivial outcomes; they are game changers of epic scale.

But can Musk succeed? Duhigg's article raises some serious questions. In addition to working insanely long hours, delivering wild tirades, and summarily terminating employees, Musk has developed a fanatical obsession with detail that puts him in the league of Steve Jobs.

At a startup, those kinds of behaviors would be unpleasant, but within acceptable parameters. When you're the CEO of a publicly traded enterprise, however, sustained erratic behavior is simply unacceptable.

From my perspective, Musk needs time to grow up. If he's not mature enough to lead a major corporation, he should hire a team of experienced senior leaders and step aside for a while. Frankly, that would be the right step. He can still play a significant role at both Tesla and SpaceX, and his rightful place in history will be assured.

Musk has already proven to the world that he is a genius. Now he needs to prove that he is also a leader.

Senior Execs Should Pay Attention to the Larger Lesson

Neil deGrasse Tyson, the celebrity astrophysicist and best-selling author, wants Musk's critics to leave him alone. Tyson compared Musk to inventor Thomas Edison in a recent interview with TMZ and defended Musk's right to smoke marijuana during a podcast with a comedian.[13]

But even Tyson concedes that Musk isn't above the law. As the CEO of a publicly traded corporation, he is required to play by the rules. There are limits to what he can and cannot say.

Musk's troubles with the U.S. Securities and Exchange Commission serve as a reminder to top-level executives in all publicly traded companies. You can and will be held accountable for what you say.

In August 2018, Musk tweeted that he might take Tesla private and suggested that he had the funding necessary for such a move. Now it seems as though he might have been exaggerating. Perhaps he wasn't thinking clearly, or maybe he hadn't gotten enough sleep. In any event, Tesla stock and Musk's reputation both took a beating.

There's a larger lesson, however. During World War II, dockworkers in the United States and Great Britain were warned that "a slip of the lip can sink a ship." The slogan was a reminder that Nazi spies had infiltrated the docks and were listening closely to gather useful bits of information that could be used against the Allies.

I thought of that slogan when I was in an airport recently and overheard three business travelers denigrating one of their clients. I was truly amazed at their carelessness and apparent lack of concern. If anyone who worked at their client's company was within earshot, the travelers easily could have put the account in serious jeopardy.

I admire Neil deGrasse Tyson and I enjoy watching his shows about science. But I think he missed a valuable opportunity to use Musk's misstep as an instructive lesson. In today's world of instantaneous communications, all of us need to exercise extra caution. We need to choose our words very carefully, especially when we speak in public or use social media.

Moreover, we need to remind our teams and our partners to exercise caution. Anyone associated with the enterprise can cause damage by speaking or writing imprudently. That's just a fact we have to live with.

Here's a suggestion: Spend an hour with your team talking about Musk and what his comments have cost Tesla. Remind them of their responsibilities and about the possible consequences of revealing company information. Invite a representative from HR and from your legal department to join the conversation. It will be time well spent.

"A slip of the lip can sink a ship" doesn't just apply to boats; it applies to the modern enterprise, too.

Notes

1. HMG Strategy, *HMG Tech News Digest*, October 26, 2018, https://hmgstrategy.com/resource-center/articles/2018/10/26/hmg-tech-news-digest-october-26

2. Harvey Nash/KPMG CIO Survey 2018, *The Transformational CIO*, 2018, https://assets.kpmg/content/dam/kpmg/xx/pdf/2018/06/harvey-nash-kpmg-cio-survey-2018.pdf

3. Robert Webb, "CIO Views on Digital Transformation: Meet the Digital CIO," n.d., https://www.i-scoop.eu/cio-views-digital-transformation-meet-digital-cio/

4. Mark van Rijmenam, "AI: A Force for Good or Bad?," *M*, June 12, 2019, https://medium.com/@markvanrijmenam/ai-a-force-for-good-or-bad-d61d7b8fd7fd

5. Capgemini, "Intelligent Automation Could Add $512 Billion to the Global Revenues of Financial Services Firms by 2020," July 12, 2018, https://www.capgemini.com/gb-en/news/intelligent-automation-could-add-512-billion-to-the-global-revenues-of-financial-services-firms-by-2020/

6. Joanna Plucinska, "European Parliament Votes to Block Copyright Reform," Politico, July 5, 2018, https://www.politico.eu/article/european-parliament-votes-to-block-copyright-reform/

7. Adam Satariano, "Tech Giants Win a Battle Over Copyright Rules in Europe," *New York Times*, July 5, 2018, https://www.nytimes.com/2018/07/05/business/eu-parliament-copyright.html

8. CFB Bots, "The Difference between Robotic Process Automation and Artificial Intelligence," April 9, 2018, https://www.cfb-bots.com/single-post/2018/04/09/The-Difference-between-Robotic-Process-Automation-and-Artificial-Intelligence

9. Dina Srinivasan, "The Antitrust Case Against Facebook: A Monopolist's Journey Towards Pervasive Surveillance in Spite of Consumers' Preference for Privacy,"

Berkeley Business Law Journal, 16 (2019): 40–101, https://scholarship.law.berkeley.edu/bblj/vol16/iss1/2/

10. Mark Zuckerberg, "A Privacy-Focused Vision for Social Networking," Facebook, March 6, 2019, https://www.facebook.com/notes/mark-zuckerberg/ a-privacy-focused-vision-for-social-networking/ 10156700570096634/

11. Ryan Holmes, "Is Content Validation the Next Growth Industry?," *Fast Company*, September 17, 2018, https:// www.fastcompany.com/90236068/is-content-validation -the-next-growth-industry.

12. Charles Duhigg, "Dr. Elon & Mr. Musk: Life Inside Tesla's Production Hell," *Wired*, November 13, 2018, https://www.wired.com/story/elon-musk-tesla-life- inside-gigafactory/

13. Tom Huddleston Jr., "Neil deGrasse Tyson: Elon Musk Is the 'Best Thing We've Had Since Thomas Edison,'" CNBC Make It, September 19, 2018, https://www.cnbc .com/2018/09/19/neil-degrasse-tyson-on-elon-musk .html.

Chapter 6

The New Customer Focus Imperative

Today's empowered customers expect and demand easy, effective, and fast experiences from brands. In fact, research from PwC reveals that 42 percent of consumers would be willing to pay more for a friendly, welcoming experience while 52 percent would pay more for a speedy and efficient experience.

Customers are also increasingly looking at a brand's experience as a differentiator as they tend to purchase experiences, not products. Because of this, the line between front-office and back-office applications is disappearing. Customer-obsessed enterprises are leveraging digital technologies to create new customer value and increase operational agility in the service of customers. As CIOs create their technology agendas (in support of their organization's business objectives), they need to take a more proactive approach to investing in technologies for competitive differentiation. These technologies should be able to bring similar experiences to customers as well as employees, which can increase return on investment and decrease total cost of ownership.

Out of the many technologies that are fueling new forms of business disruption, the maturity and evolution across the following areas provide CIOs the capabilities to create compelling experiences and generate operational gains:

- **Hosted cloud infrastructure to native.** Flexible and modern cloud architectures built on containerized microservices architecture will be able to provision machine learning compute speed faster than legacy-hosted cloud infrastructure. They are able to provide a framework to evolve with the changing behavior of the user that keeps evolving while lowering the operational cost to run these systems. The new cloud infrastructure provides the flexibility and agility of iterative change in this new age of IT.

- **Chatbots to conversational AI.** Conversational interfaces are becoming the new definition of self-service and personalized experience. This represents an experience that is injected at the point of entry either for your customer or your employee engaging or for interacting with a system. Here you will start to see chatbots moving from manual rule-based systems to decision-making systems that are evolving with each interaction (supervised to unsupervised learning). The true differentiator will be choosing a tool that is easy to use and maintain and built without admin intervention. Conversational interfaces will add more value by increasing customer satisfaction and net promoter scores.

- **RPA to autonomics.** RPA traditionally was used to automate specific tasks, especially business processes. Autonomics are being designed specifically for IT-related functions and processes in the areas of server, network, and application management; database administration; virtual machine provisioning; and diagnostics. These systems have the ability to learn new capabilities and respond to new conditions and reduce man-hours required to answer repetitive requests.

By investing in these technologies and building capabilities that can support their businesses, CIOs are able to show a strong and positive ROI by directly impacting the brand experience and reducing operational overhead while serving both external and internal audiences.

Rejuvenating the Customer Experience Through AI

CIOs are under increased pressure to help their organizations improve the customer experience. This makes perfect sense, given CIOs' unique view across the enterprise and how they are able to connect the dots on how people, processes, and technology can be blended to deliver exceptional customer experiences. It is little surprise then that 65 percent of "digital vanguard" CIOs report they have strong relationships with customer-facing business functions, according to Deloitte's 2018 Global CIO Survey.[1]

As part of this, CIOs should work with customer-facing functions (sales, marketing, customer support) to invest in AI and digital transformation to improve user experience and customer satisfaction and to run the business smoothly without disruptions and outages.

There are many areas that all companies can invest in to improve the overall customer experience (CX). Each organization needs to make bets on which AI technologies offer short-term and long-term ROI while improving the customer experience.

CIOs are under constant pressure to improve user experience and deliver more business-aligned value through technology and best practices. CIOs are also under pressure to reduce both operating expense and capital expenditure costs while drastically improving customer experience. CX is a key ingredient for organizations to improve user experience and customer satisfaction. Customer experience must include wide-ranging areas, such as user engagement, user retention, growing the user base/community, increasing user adoption, and providing customer services platforms for users to innovate. Although this is an area that has been perfected by consumer-facing companies, even traditional enterprises are now aiming for the same goals.

CX can be improved by automating tasks in three ways: intelligent automation and RPA; intelligent dialog/conversation AI; and AI-driven service management as a service. The use of AI in all these areas aims at creating a very dynamic, personalized experience that minimizes the mean time to resolve customers' requests.

The following areas are important for CIOs to consider for overall CX improvements:

- User behavior analysis to provide personalized experiences

- Proactive customer service to users

- Predictive customer support to better anticipate and respond to customer issues ahead of customer outreach

- Prescriptive solutions to resolve customer challenges

Organizations of all sizes can now easily deploy intelligent automation technologies that fuse RPA with cloud-native orchestration and AI and transform traditional service desks, help desks, and cloud/IT operations. These digital transformations can be accomplished independent of legacy IT organizations, legacy IT service management solutions, or outsourcing service providers. One way this type of IT service management (ITSM) automation can occur is by utilizing multicloud, robotic process automation (RPA), virtual agents, artificial intelligence for IT operations (AIOps), and AI-driven service management (AISM).

The key solutions and technologies where CIOs can invest for improving customer experience can be categorized as follows:

- Next-generation ITSM (AISM)

- AIOps for cloud/IT/DevOps

- Intelligent automation or RPA

- Automated service management—virtual agent and agent intelligence

Here are some of the wide-ranging benefits of AI-enabled CX to users and customers:

- Awesome user experience that increases the overall customer satisfaction

- Fast resolution of customer issues and self-service capabilities

- Drastic reduction in outages and improved business uptime

- Significant reduction in mean time to repair during outages

With AI and machine learning for customer experience improvement, and by automating repetitive and programmatic tasks, customer service teams can channel their creativity, passion, and imagination into actions that provide greater value to the dynamic and growing business while also achieving drastic operational cost savings in the first year.

Reimagining the Restaurant Guest Experience

Customers expect high-quality experiences in every touch-point they use with the companies they do business with. The restaurant guest experience is no exception.

As Deloitte pointed out in its "Restaurant of the Future" report, "winning restaurants will be the ones that invest wisely in digital, operations, marketing, and technology, and can harness the power of their employees to serve as brand ambassadors at the moments that matter."[2]

Customers are increasingly expecting restaurants to deliver on these capabilities. For instance, 31 percent of consumers surveyed by Deloitte want the ability to pay their restaurant check with their smartphones so they don't have to wait for a server to deliver their bill and then process it.

Donagh Herlihy is tapped into delivering on these height-ened customer expectations. As EVP and CTO at Bloomin' Brands, which operates casual dining chains such as Out-back Steakhouse, Bonefish Grill, and Carrabba's Italian Grill, as well as the fine dining concept Fleming's Prime Steakhouse & Wine Bar, Donagh and his IT team are focused on how they can help the company to reimagine and reinvent the guest experience at the company's restaurants.

I sat down recently with Donagh to talk through his per-spectives on technology leadership and discuss how the IT organization can help Bloomin' Brands to be more attuned to its customers' needs and preferences. Here is a lightly edited transcript of our conversation:

Hunter Muller: What are your focus areas today?

Donagh Herlihy: With over 1,400 locations, the first area we focus on is in elevating the customer experience inside our restaurants. This means leveraging technology to delight our guests. As full-service restaurants, we seek to do this unobtrusively. However, whether it is in enabling you to get seated quickly, ensuring that your food is prepared exactly as you requested, or enabling you to pay and go when you want to, we are constantly seeking to remove any friction in the guest experience.

The second focus is on our transformation to an omnichannel business model. We are moving to a world where a significant part of our business will be in delivering meals to you, wherever you are. As consumers, we are accustomed to this level of convenience in categories like pizza with brands like Domino's. This is a growth opportunity for casual dining brands and we want to lead the way.

We started our delivery business with our own driver fleet in 2016, and we have been progressively fine-tuning our capabilities and rolling it out across our stores. We anticipate this "off premise" dining opportunity growing in the relatively near term to about 30 percent of our business. To enable this, we have built out a platform of proprietary e-commerce and logistics technologies that enable easy online ordering and payment as well as rapid delivery.

The third focus area is in the realm of generating increased traffic through our loyalty program. Also launched in 2016, we now have over 8 million guests in the program, and it is contributing a measurable lift in our sales. However, the real value in the program for us is in using the detailed data about our guests and their preferences to personalize our communications to them and their experiences with us. We are now using machine learning to customize our direct marketing and our digital campaigns. As part of this, we are using customer data to better understand our customers and their behaviors and preferences.

HM: What are some other steps you're taking to improve customer experience?

DH: Most of our brands are casual dining brands where you don't make a reservation. Guests do not enjoy arriving at a restaurant and having to wait for a table. So, we expose in real time on our websites and mobile apps the wait times for each of our locations so that a guest can put their name on the list and arrive as the table is ready. The technology creates transparency at scale and eliminates this point of friction.

Guests like the convenience of paying their restaurant bills by phone so they don't have to wait for their servers to deliver and process their checks. So, our mobile apps provide for payment as well as for tracking and redeeming loyalty program rewards. Providing these capabilities reduces customer anxiety.

We're also optimizing orders for food delivery so that an order reaches the customer within ten minutes of leaving the kitchen. The logistics involves hitting a very precise window to have a driver available to take that order and deliver it quickly. We're using cloud technologies to build, test, and learn based on feedback from our customers and our operators. We now have 900 restaurants delivering this convenience to our customers.

Looking ahead, we want to use technology to determine whether a customer entering one of our restaurants is a loyalty member, to give them an appropriate welcome and to identify their preferences proactively. That's bringing personalization to life in the store.

These and other customer-facing solutions we work on are all derived by listening to our customers and by understanding the customer journey with each of our brands.

Behind the scenes, we have done a lot to enable the business performance with technology. With roughly 90,000 employees, it is critical that labor scheduling and utilization in each store and on each shift is optimized. This is also the case with food and ingredients. We are using cloud-based platforms to mine data and predict the best food and labor allocations to each and every shift, and this is critical in maintaining our value proposition in a world where food costs and labor rates are inflationary. We want to use data and algorithms to avoid passing on cost increases to our guests.

HM: What are some of the results so far?

DH: In terms of food and labor, we have taken tens of millions of dollars of cost out of the business.

In terms of guest traffic and sales, the loyalty program and the delivery program have been the two biggest growth levers we have enjoyed over the past couple of years. I cannot go into specific figures, but both are meaningful in terms of profitable growth.

Inside IT, we have focused on strengthening our culture and our processes and have been moving progressively to a fully Agile development and deployment process. Our IT organization started with this two years ago, and our CEO began promoting this challenge to the rest of the company in late 2017. This is enabling us to work with a different mind-set, to be more open and collaborative, to take risks and to test and learn faster. We're making good progress there.

HM: What's the size of your IT team?

DH: Our IT staff is relatively small, with about 160 people internally and 70 to 80 contractors and partners.

HM: How would you characterize your role in redefining and redesigning the company?

DH: Speed of decision making and a willingness to take measured risk are characteristics that I feel strongly about promoting. Many decisions involve multiple functions or brands and can get bogged down seeking consensus. In our Agile transformation within IT, we created the mantra: #SafeToTry. In those cases where we don't all agree on a decision, we reframe the question from "Can we all agree?" to "Is it safe to try?" Many times a decision or an idea is safe to try, and it is better to try and learn and pivot than to engage in paralysis by analysis.

HM: How would you characterize your leadership and collaboration style with fellow members of the C-suite?

DH: One thing that I've always done is seek to have empathy for my peers, their issues, their agenda, their world. I want to walk in their shoes to the greatest extent possible, and then I can be a better partner to help them and the enterprise be successful.

An avenue into that is to spend time on the frontlines. I spent my first month at Bloomin' Brands training in the kitchens and in the restaurants to learn the operational day-to-day reality faced by chefs, kitchen staff, servers, and managers. That sent a message to my people about where our focus needs to be. It also gave me some street cred with the senior leadership of each of the brands in that I was coming to them from a place of business operations and not from the point of view of the CTO. They appreciated that I cared about how the technology was being used in the operations and that I was willing to dig in with hourly staff to experience their world. This helped me to ask better questions: Why do we do certain things? Why are these processes priorities?

HM: You led the creation of the company's consumer digital roadmap. Tell us about that and how you partnered with the CMO and other members of the executive team in bringing this to fruition.

DH: When I got here in late 2014, we put together a small cross-functional digital innovation team. We had someone who had a digital consulting background, an ex-CMO for one of our brands, and a couple of digital product managers. We worked with our consumer insight folks to stage a series of focus groups in our restaurants with a cross-section of our guests, with all relevant demographic and psychographic profiles to understand and map out their customer journeys.

Those focus groups began to align us on understanding the customer—When does she begin thinking about eating lunch or dinner? What tools and resources does she use to make her decision about what and where to eat? How do we engage with her in that journey? Then we worked with those insights and validated it through broader online consumer research to help us to prioritize pain points and potential solutions. We isolated about a dozen pain points, such as customers feeling they were being held hostage waiting for the check, and then were able to work through the operational and technology changes that were needed.

We saw new opportunities, too, such as seeing the pent-up demand for delivery in casual dining, enabling us to address an unmet need. We were able to assess the size of the prize in early tests through third-party delivery platforms such as Grubhub. Our CEO was out in front of this, which enabled us to build out a small team with a lot of autonomy to test this delivery with our own staff at a few restaurants, first as a phone-ordering business only, and then we fast-followed with our e-commerce platform. And it grew from there.

CIO's Role in Driving the Business Forward

CIOs play an increasingly strategic role in driving innovation and business transformation initiatives. This helps explain why 79 percent of CIOs who were recently surveyed by Grant Thornton and the Technology Business Management Council say that IT has a voice in business strategy and business initiatives.[3]

Of course, competitive pressures come into play as well. "Amazon has forced us to become nimbler and more responsive," says Prabhash Shrestha, EVP and chief digital strategy officer for the Independent Community Bankers of America.

Several CIOs and technology executives said they are using various technologies within their organizations to cater more effectively to customers and drive higher levels of productivity.

"From a technology perspective, we are aggressively pursuing a digital service experience that complements our face-to-face client coaching relationships with our advisors. The client experience is central to the technology transformation," says Amy Doherty, EVP and COO at First Command Financial Services, Inc. "Our Straight Through Processing capability automates the entire transaction component of the financial planning process for clients, creating a customized paperless experience with higher quality, allowing our clients to focus on their financial goals and coaching relationship instead of filling in repetitive forms."

In other cases, technology executives and their teams are leveraging technology to create added value for customers. "We're using machine learning to combine satellite data from NOAA with our own data to provide value to our customers," says Curtis Generous, CTO and VP of engineering at Earth Networks.

Notes

1. Deloitte Insights, "Manifesting Legacy: Looking Beyond the Digital Era, 2018 Global CIO Survey," 2018. https:// www2.deloitte.com/content/dam/insights/us/articles/ 4774_CIO-survey/DI_CIO-survey-2018.pdf

2. Deloitte, "The Restaurant of the Future: Creating a Winning Guest Experience," n.d., https:// www2.deloitte.com/us/en/pages/consumer-business/ articles/restaurant-future-survey-technology-customer-experience.html

3. Grant Thornton LLP, "Survey: CIOs See IT Moving from Cost Center to Trust Center, Even as Challenges Abound," February 26, 2019, https://www.grantthornton .com/library/press-releases/2019/february/CIOs-see-IT-moving-from-cost-center-to-trust-center.aspx

Chapter 7

Next-Generation Leadership

As enterprise companies continue to take more active steps in increasing the diversity of their leadership teams, there's been a steady rise in the number of racially diverse CIOs and women taking on leadership roles.

Although the percentage of female CIOs in Fortune 500 companies has risen from 15.6 percent in 2014 to 17 percent in 2016, according to SpencerStuart's The State of the CIO in 2018 study, women are still vastly underrepresented both in IT and in IT leadership roles.[1]

I recently spoke with Phyllis Post, former vice president and CIO, Global Human Health IT, Merck, to discuss steps that could be taken to attract more women into IT careers and to strengthen career opportunities for women in the technology field. Here's a lightly edited transcript of our conversation:

Hunter Muller: Your career originally began outside of IT. Can you please share how your entry into an IT role came about and what this may portend for other women who may be interested in launching a career in IT from another profession?

Phyllis Post: My career began in the typesetting and printing industry in a period when it had transitioned from hot type to cold type where we had big rooms with temperature controls for large mainframe systems and you did your typesetting there. A few years later, desktop publishing emerged and began to change the industry—ultimately leading to a total disruption in its future.

During that period of change, the typesetting and printing industry was becoming more digitally enabled. Although I was on the business side of the organization, enabling the company to change to that early version of digital was part of my role. So, I helped make that transition from those larger mainframe systems to desktop systems, eventually leading to my heading up the web division for one of the companies.

It became clear in the mid-1990s that the printing industry was changing significantly, including a lot of consolidation. I was looking for the appropriate next steps that would not only provide stability for professional growth but also a mechanism that allowed me to contribute. I was approached by a recruiter on behalf of Merck about an IT project manager/planning role based on my background. I knew going into pharma would be a huge learning opportunity and that going into IT was going to be a much different environment.

That major switch was both frightening and exciting—and set the context for my career. One of the approaches I took when I first joined was really immersing myself into what I needed to learn and doing research to understand the industry and where it was going both for pharma and in terms of the technology being used. One of the things that helped me to make that transition was learning to code.

Although I was extremely nervous in making that change in career direction, the approach I took has helped me when I've needed to work on other new or transformation initiatives at Merck. And I've consistently found that if you apply yourself to learning something new, you can be very successful in taking on new challenges or moving into an entirely new space. In fact, often you can contribute more from not having preconceptions about a subject.

In terms of women exploring roles in IT—we as women need to be more confident that even if we don't know a subject or have experience in a particular area, we can learn it. You just need to be willing to face that fear and go after what you want. If you have a growth and learning mind-set, you can do it.

While the functional skills I needed to learn are different now, the leadership skills have remained consistent—building strong relationships across borders and boundaries, helping to articulate and connect the dots for others, setting a North Star aspiration, inspiring both individuals and the team, and continually focusing on talent development and growth. Honing those elements and having that strong leadership core is something that other women can also focus on.

Strong collaboration and communication skills and the ability to help people work together across borders and boundaries are things that can be applied in any other environment.

HM: What kinds of opportunities did you take advantage of during your career that led to your current role as VP and CIO, Global Human Health IT? Can you also offer some suggestions for female IT professionals who want to grow into IT leadership roles?

PP: Since I've been at Merck, I've had the opportunity to grow and change roles quite a bit. I have moved across the various divisions, worked in roles that were globally oriented, and had an expatriate opportunity to work in Asia. Having these varied and disparate growth opportunities has kept it fresh and made it seem at times as if I were working for different companies.

Although I've had a North Star for where I wanted to go professionally, I was also willing to take on assignments that were either outside of my comfort zone or allowed me to enhance the breadth and depth of my knowledge. Not just across a specific area of the business but across the entire value chain at Merck, from research to product commercialization. It helped me to envision the outcomes of the business as well as the network of connections to drive toward those goals.

Having that breadth of knowledge allows you to look at challenges or opportunities differently. You notice things that you may have otherwise missed since you're looking at it with a broader lens.

One thing I tell female technology professionals when I meet with them is that sometimes we can be hesitant to go outside our comfort zones, and that as women leaders we should not be afraid to take chances. You don't know if you don't try, and you don't want to regret playing it too safe.

For both women and men, it is also really important to figure out what your professional aspiration is and actively work toward that goal—including being a vocal advocate for yourself. You need to share your goals with your managers and mentors and proactively see how different opportunities might help bring you closer to your goal by providing the right experiences and skills.

HM: From your perspective, what can be done differently in this regard that could help to increase the percentage of women in technology leadership roles?

PP: We need to help emerging female leaders and encourage them to take chances on new opportunities that are outside their comfort zones or that are a stretch but that can help them grow and move toward their professional goals. They also need to actively advocate for themselves—both for their aspirations and where they want to take advantage of new opportunities. Men tend to be more comfortable with those types of conversations. We can help female leaders get more comfortable with these discussions and approach through mentorship, coaching, and additional education around professional branding.

HM: What are some ways we can better engage young women in IT and STEM career paths at an early age?

PP: We need to do more with children in middle school before they get to high school to get them interested in these fields. We need to provide them with a better understanding of the types of opportunities that are available. We can do this by showing them examples of successful female leaders, whether it's in highly scientific or engineering areas or in IT. We can point out role models for them to look up to, proactively encourage young women to get involved in STEM activities early, and persuade them not to lose confidence if they run into any challenges along the way.

We also need to continue to focus on the corporate messaging and actions around diversity, especially around diversity of thought and inclusion. This will help encourage young leaders that, even if they don't see it right now, there are increasing opportunities open to them and that things are changing.

HM: How useful are current programs for young women, such as the Girl Scouts or STEM camps for girls?

PP: Girl Scouts, Girls Who Code ... these types of programs that are targeted at encouraging young girls are extremely useful. It helps to give them confidence and to provide opportunities to develop these skills in a safe environment. It's important to nurture these opportunities.

HM: Research reveals that a higher percentage of women opt out of tech careers than their male counterparts. In fact, some studies suggest that more than twice as many women leave the technology industry as men. What are some steps that could be taken to help encourage women to remain in technology roles?

PP: Before women even enter the industry, they may get frustrated or nervous about what they hear about the industry and how much growth they can have based on what they see in the real world. We need to actively promote diversity and inclusion in our organizations to help make young women comfortable.

Also, if they see their male counterparts getting opportunities or promotions more frequently than they do, they can become frustrated. It is important for leaders to coach them—encourage them to look at different opportunities, to take on a role outside of their comfort zone, and to help them get more comfortable in advocating for their future.

I've also heard some women express concern that there's an expectation for them to make a choice between family and a career in tech. Integration of both is possible. We need to work at correcting this misperception, or where there may be an issue, women need to advocate for the balance that works for them.

HM: What else can be done?

PP: We need to deliberately encourage women to come into STEM occupations. We also need to continue to coach and mentor young women as they come into the industry so that they remain and continue to grow. And wherever possible, we need to actively point to female role models that will give them the confidence that they can successfully make the journey.

Workforce Diversity at All Levels

Companies with diverse workforces—in terms of gender, race, ethnicity, sexual orientation, and socioeconomic backgrounds—enjoy multiple benefits. A *Harvard Business Review* study has found that cognitively diverse teams solve problems faster than teams of similar people.[2]

Meanwhile, a January 2018 study conducted by North Carolina State University's Poole College of Management revealed that companies with policies that encourage the promotion and retention of a diverse workforce perform better at developing innovative products and services.[3]

These characteristics play out across all levels of the enterprise, including within the IT organization. I spoke recently with Earl Newsome, CIO, Americas IT, Linde PLC, and former global CIO and VP at Praxair, to get his views on the benefits of developing diverse teams along with his recommendations for developing diverse teams. Here's a lightly edited transcript of our dialogue:

Hunter Muller: Why is it so important for organizations to have teams from diverse backgrounds?

Earl Newsome: Ultimately, your best and most innovative solutions come from examining the widest variety of views and opinions. Only through transparent dialogue that examines different perspectives and vantage points can you establish what's best for your organization.

It's not dissimilar to casting a play or a performance. Each character has a unique role and if you fill it with the same exact type of person, the result would be a pretty boring show. But a vibrant and diverse cast brings something unique to the performance, resulting in a more dynamic and successful outcome.

HM: What are some of the ways you had approached this at Praxair?

EN: It's important to understand there are many elements of diversity you can assess, such as background, education, race, religion, and country of origin. I always try to make sure I have these various elements in my environment, both inside and outside the company. With regards to inside the company, I work hand in hand with our corporate diversity management and talent acquisition teams to ensure I have the best talent and interns that's available in a marketplace that is rapidly diversifying every day.

But it's not enough to simply seek diversity. You must ensure these varying perspectives are being listened to as well. Diversity is the measure of differences you include on your team, but inclusion is a choice, and as a leader you must make sure you're seeking and utilizing all those different views and perspectives.

The last element is collaboration. That's where the rubber hits the road. You have to take all those different ideas, suggestions, and points of view and work collaboratively to incorporate them into your strategy.

One example is to focus on your intake through recruiting initiatives. Teams must go to the places where they can find the best talent and ensure we were very inclusive in sourcing applicants. First and foremost, our goal was to fill our organization with the very best talent. This includes obtaining the best and most varied perspectives.

HM: What are some of the lessons you had learned over the course of your career that you were able to apply in your approach to strengthening diversity and recruitment at Praxair?

EN: The first is, you need to seek out the dissenting opinions in the room. Ask for them. Usually that dissenting opinion is there, it just may not have been expressed. The person may be an introvert or is concerned about saying the wrong thing. You need to make sure you draw out all those ideas.

Second, you have to get your passport stamped. In other words, you have to go out there and meet the people.

You can't just sit in your office. I often bring in ten random people, offer coffee, and connect with them. I do this when I travel, when I meet with customers and with my team members. I want to learn about what's going well, what keeps them up at night, and what's on their wish lists. This empowers people to speak up and helps you get all those diverse perspectives from across your business.

Finally, a great way to drive diversity is to give thanks. As people within your organization are executing on the strategy, be sure to provide them with rewards and recognition. Recognize the people in your organization and the value that they're delivering. One of the ways I do this is with thank-you cards.

HM: What have been some of the results?

EN: I think what you see and the reward that you get is buy-in. I don't know if you can quantify buy-in, but if you're not seeing the kind of progress you expect, you're not getting your message across.

HM: What are some recommendations you'd offer to your peers on things they should be thinking about and doing with regards to diversity?

EN: Reach out to people at all levels across your organization and let them know you want to hear their ideas. That alone will help people feel included.

Then take that a step further by carrying out little acts of inclusion. For instance, take someone to lunch you might not otherwise connect with. You may be surprised what you'll learn, and that individual will likely feel more appreciated and included.

You may also want to consider offering unconscious bias training and starting employee-run resource groups for female and diverse employees. Always remember, for an organization to reach its full potential, all team members must commit to building a diverse and inclusive workforce where all views are respected and heard.

Diversity as an Economic Imperative

Executives often talk about how diversity of leadership and organizational teams can result in greater innovation. The concept that diverse teams across gender, race, ethnic background, religion, education, and other characteristics will generate an abundance of different perspectives and ideas that can help fuel innovation seems logical. And there's also research to back this concept.

The study conducted by North Carolina State's Poole College of Management mentioned earlier evaluated diversity policy data on the 3,000 largest publicly held companies in the United States, along with patent data and new product announcement data from 2001 to 2014.[4] The findings revealed that "a company that checks all of the diversity boxes" saw roughly two new additional product announcements over a ten-year period.

"Given that most firms produce an average of two new product announcements per year, that's significant," said

Richard Warr, coauthor of a paper on the study and head of the Department of Business Management at the school.[5]

For a deeper dive into this subject, HMG Strategy spoke with three executives: Zackarie Lemelle, president/CEO, New Hope Coaching and Consulting; Beverly Lieberman, president, Halbrecht Lieberman Associates; and Lynn McMahon, managing director and North America Media and Entertainment lead, Accenture. Here is a lightly edited version of our conversation:

HMG Strategy: Why is it critical for executives to foster a culture of diversity and inclusion to help strengthen innovation?

Zackarie Lemelle: It's been proven time and time again that diversity and inclusion are absolutely critical to innovation. You cannot innovate with people of a like mind. Different ideas can be coalesced to drive innovation. We all come with different ideas and different perspectives.

To quote psychologist Kenneth Kaye, "If we manage conflict constructively, we harness its energy for creativity and development."

Beverly Lieberman: If you look at the workforce here in America, there's been an influx of software and electrical engineering talent that comes from overseas. U.S. homegrowns are no longer dominating the tech workforce.

For years, we liked to hire in our own image, and that got us through the 20th century pretty well.

But changes in demographics in the workforce and an influx of non-Americans have fostered a need to embrace diversity.

Managing a multicultural workforce takes skill. Companies like Johnson & Johnson have done a lot of diversity training with their leaders to bring out the best in people that don't necessarily come from the same culture. They need diverse leadership to reflect the diversity of their customers.

Lynn McMahon: I once heard Merck CEO Kenneth Frazier speak at an event. He was asked about the business case for diversity, and his response was "I've never heard the case against diversity."

At Accenture, we see our clients and our business undergoing a lot of disruption, as we move from a product world to a digital and services world. In order to make that transition, culture is important, and culture is fed by innovation.

Diversity is a key part of this. We believe it's very important for people to bring their best selves to work, and diversity is a critical part of fostering this. Diversity and innovative thinking don't occur overnight. You have to create opportunities for people who think differently. We've continued to take actions that enable us to attract, retain, and grow people—both within our company and with our clients.

To do that, you need to have transparency about diversity initiatives. Diversity has historically been set behind the curtain to some extent. For instance, we've made a commitment to bring ourselves to a gender 50/50 work-force by 2025 on a global basis. In some markets, we're already there.

HMG: How can a culture of diversity and inclusion can help to attract talent?

BL: It's certainly easier for me when I'm recruiting to be able to say that 20 percent of the leadership team is Indian or female and would welcome you as a member of the leadership team. It's not just a White Protestant culture here. Demonstrating that a company's culture is inclusive helps to make people feel comfortable.

Twenty-five years ago, women brought onto a senior leadership team were often the only women on those teams. Most didn't want to be groundbreakers. It was imperfect, and it still is at many companies.

If you've got a little bit of diversity and are trying to attract diverse talent, you're likely going to stand out among your competitors. It just makes good business sense.

LM: What I have seen, especially with the Millennial generation, is that they're not just looking for a job and a paycheck but a place where they can contribute. The goals and ethics of a company matter a lot to people.

The things that we do both inside and outside of our organizations help to market our diversity initiatives. I'm the executive sponsor for interfaith initiatives at Accenture. As a company, we're actively involved in Women in Tech NY, which is aimed at getting more women and female minorities into computer science disciplines.

ZL: When I started in IT 50 years ago, there were no African Americans. When I got into the field, there were no people of color.

Then you began to see diversity over time.

If you want to bring me into the fold, then you have to make me feel like I have a place where I am valued, where I matter. If I don't see anyone who looks like me, then I will question whether and where I fit in.

HMG: What should be the CIO's role in addressing diversity?

LM: It's important to have flexibility in the not-one-size-fits-all approach. Women who go out on maternity leave should have different options for reentering the workforce.

As we've attempted to try to get more women into STEM, they tend to think of technology professionals as guys wearing hoodies working in a basement. What we're trying to describe to them is the ability to help work on world problems that women care about.

CIOs who are visibly active on diversity issues are hugely important. Also, it's important to demonstrate they are building a succession plan, a recruitment plan, and a training plan that encourages a diverse workforce.

ZL: When I was a CIO, I always wanted to tap into the greatness of all of my people regardless of who they were, what their backgrounds were, or where they were located. This is where the best ideas and energy come from.

When you're excited about something and engaged, you automatically go the extra mile and tackle that extra detail. Because you feel that you matter.

Without high engagement, you don't get innovation.

BL: The CIO is in a position where she or he plays a role in bringing in talent. A lot of the rank and file are programmers, analysts, data scientists, architects, network engineers, etc. Thirty to forty percent of those candidates come from diverse backgrounds.

It's one thing to attract diverse candidates, it's another thing to retain them. The CIO should partner with HR and bring best practices into the IT organization for sensitizing managers on what needs to be done to work with people from diverse backgrounds as well as communications and training.

CIOs should also help advance the message that we will do better as an organization if we treat diverse candidates with dignity.

Notes

1. SpencerStuart, "The State of the CIO in 2018: A Three-Year Study of a Rapidly Changing Role," February 2018, https://www.spencerstuart.com/research-and-insight/the-state-of-the-cio-in-2018

2. Alison Reynolds and David Lewis, "Teams Solve Problems Faster When They're More Cognitively Diverse," *Harvard Business Review*, March 30, 2017, https://

hbr.org/2017/03/teams-solve-problems-faster-when-theyre-more-cognitively-diverse

3. Matt Shipman, "Study Finds Diversity Boosts Innovation in U.S. Companies," *NC State University News*, January 9, 2018, https://news.ncsu.edu/2018/01/diversity-boosts-innovation-2018/

4. Ibid.

5. Ibid.

Chapter 8

The Future Is Already Here

As the dawn of quantum computing draws nearer, it's fair to ask whether transformational technology leaders should begin studying how quantum computers work.

From my perspective, the short answer is yes—even if the CEO and executive board is unlikely to start quizzing you on the topic of quantum computing anytime soon.

There's a simple reason you should care about quantum computing: It's going to completely transform the IT universe. When quantum computers become operational, they will surpass traditional computing systems almost immediately.

"Computer scientists Ran Raz and Avishay Tal provide strong evidence that quantum computers possess a computing capacity beyond anything classical computers could ever achieve," Kevin Hartnett wrote in *Quanta* magazine.[1]

In business, quantum computing would translate into a broad array of competitive advantages, including faster design, testing, and deployment of new products and services. Quantum computing is expected to revolutionize scientific research, leading to rapid advances in medicine, manufacturing, logistics, marketing, and financial services. Anything you can do with a traditional computer you'll be able to do faster with a quantum computer.

But that's just where the advantages begin. With quantum computing, you'll be able to tackle a wider range of complex challenges, such as forecasting, predictive maintenance, automation, robotics, and AI development.

We'll discover new medical therapies faster and invent materials that are lighter, stronger, and more flexible than anything we have today. With quantum computing, we're likely to live longer and more productive lives.

"Theoretical computer scientists already knew that quantum computers can solve any problems that classical computers can. And engineers are still struggling to build a useful quantum machine." But according to Hartnett, a paper that Raz and Tal posted online "demonstrates that quantum and classical computers really are a category apart—that even in a world where classical computers succeed beyond all realistic dreams, quantum computers would still stand beyond them."

I believe quantum computing will become a driving force for change and innovation. It will profoundly transform our lives and the lives of our descendants. I predict the mid-21st century will be remembered as the time when quantum computing began reshaping the world.

As a transformational executive, you are expected to lead, reimagine, and reinvent. Quantum computing will become a standard component of the IT portfolio. Now is the time to begin learning more.

Mandate for Disruptive Transformation

Two-thirds of C-suite executives believe that four out of ten Fortune 500 companies will no longer exist in ten years as a result of digital disruption, according to a survey of more than 500 C-level executives conducted across the United States and Europe by ChristianSteven Software.[2]

In order to help their companies to survive and thrive, CIOs and other organizational leaders must reshape their leadership philosophy. Leaders in these highly disruptive times must think outside the whitespace to drive innovation for the business.

Today's leadership refresh includes being proactive about new business opportunities and transformational technologies to keep challenging yourself and staying ahead of competitors. "FAANG-BAT companies such as Amazon and Google are elevating customer expectations," says Sangy

Vatsa, EVP and CIO, Comerica Bank. We as organizational leaders have to get out in front of it to lead the digital change for our business, anticipate customers' growing expectations and deliver to it."

One of the ways that technology executives can think differently is by examining the principles and practices of successful companies such as Amazon to inspire creative ideas for their own organizations. "Amazon's first leadership principle is being customer-obsessed," says John Rossman, former Amazon executive, managing partner, Rossman Partners, and author of *The Amazon Way Series*. "It's not just a headline, they truly mean it. This includes drawing off its Voice of the Customer program to better understand and act on customer needs and interests."

Digital-era technologies also require having a digital-era mind-set. According to Christopher Mandelaris, VP, chief information security and privacy officer, Chemical Bank, "You can't think linear/analog; you have to think digital/exponential to succeed in today's competitive environment."

It's also imperative for executives to cultivate a culture of innovation. This includes providing employees with the time and resources to brainstorm on innovation concepts together. At Ally, "we give people time off their day jobs to connect with other people and drive innovation," explains Angie Tuglus, COO, Ally Insurance, Inc.

Art of Effective Communications

Reawakened leadership also demands that leaders improve their listening and communications skills. "To listen and communicate effectively, seek to understand and recognize that your viewpoint is the least important at the table," says Renee Arrington, president & COO, Pearson Partners International, Inc.

Indeed, according to Patricia Watters, managing partner, DHR International, communication "is often what is lacking in most CIO turnover situations."

Thanks to their unique view across the enterprise and how people, processes, and technologies are conjoined, CIOs and technology executives can identify and suggest opportunities for transforming the business.

"At the end of the day, innovation is about business survival," says Vijay Sankaran, CIO, TD Ameritrade.

"You want to constructively disrupt your own business," Sangy Vatsa adds. "Because if you don't, someone else will."

Creating Value Continually

CEOs fully expect CIOs and technology executives to help them to identify and execute on new business opportunities. Doing this includes the ability to leverage organizational data for competitive advantage along with the ability to enable innovation on a consistent basis.

But the CIO's role doesn't stop there. CIOs also must help the CEO, CMO, and line-of-business leaders to cater to changing customer preferences and needs.

Vish Narendra, SVP and CIO, Graphic Packaging Corporation International, says, "The distinction between IT and the business has disappeared; IT is the business. "The tech leader for an organization should be well versed in all parts of the company and be able to work with the functional and P&L leaders to identify their needs."

Vish and other technology executives shared their insights on the role that the CIO can—and should—play in helping fellow business leaders to identify and execute on new business opportunities and in helping to cultivate a culture of customer centricity.

"Most of the CIO searches we've been conducting over the past few years have been focused on finding executives who can deliver on heightened leadership expectations for the organization," says Steve Kendrick, president of KER Partners. "Part of this has been driven by the prior CIO in place not delivering that level of leadership, including clearly defining what digital is and what it represents for the organization."

These leadership capabilities include the ability to understand rising customer expectations from the customer's perspective. "It is absolutely imperative that the next-generation

CIO is looking at it from the lens of the end customer," says Eric Anderson, partner and global head of the Tech Enabled Services Practice at Egon Zehnder. "CEOs are much more vocal about what they want from CIOs. If a person today is just keeping the lights on and can only incrementally participate in strategic discussions, then they're not the right fit. They're not positioned organizationally to be effective."

"Within our business, the market is undergoing a seismic shift and technology is a key catalyst for us in that journey," adds Gautam Vyas, EVP and division executive at FIS Global, a global provider of technology to the financial services industry. "The CIO needs to act as a partner and an enabler. If the business needs to launch a new service, you need to have the right capabilities in place to respond to an opportunity that's imminent or on the horizon. And the technology selected must be scalable."

"What's really been top of mind for next-gen CIOs is that they're always listening and always learning," says Dr. Kenneth Russell, former VP, Digital Transformation, and CIO at Pfeiffer University. "It's about being a valued voice in the C-suite. I've been in the CIO seat several times in my career and what's different this time is that I'm able to have meaningful conversations with the other CXOs—I'm not just the IT guy."

Too often, companies can get bogged down by core systems that take an inward-looking view. As a result, this positions the company to operate as to how it wants the customer to behave, says Vish Narendra. Ultimately, this results in customer frustration.

"In these instances, what the IT team can do is to help sales, marketing, and customer service provide a 360-degree view of the customer that expands all boundaries," Vish explains. "This can help improve customer satisfaction and customer retention while mutually benefiting both the customer and the company in terms of its profit margins."

In addition, the CIO can work with the company's chief customer officer or chief revenue officer and other customer-facing leaders to better understand how the customer experience is being captured and measured across each channel, Steve Kendrick says.

"Does that mean the company should strive to become an Amazon or another category leader? Not necessarily. But there are likely best practices that can be gleaned and applied," Steve adds.

Welcome to the Era of Trillion-Dollar Market Caps

Two major events in the technology world occurred in 2018. One generated headlines in every major media outlet while the other was barely noticed.

The event that caught everyone's attention was Apple's market cap reaching $1 trillion. Apple became the first publicly traded U.S. company to hit the trillion-dollar mark, and now it seems likely that Amazon and Google will soon achieve or possibly surpass Apple's phenomenal valuation.

Without question, we have entered a new era of wealth and value creation. But these new levels of wealth and value also pose risks. In the past, hundreds of companies contributed to the economic health and well-being of our nation. Today, the list of supercompanies has dwindled significantly.

In other words, fewer companies are creating more wealth and value than ever before. We've all heard the old saying, "Don't put all of your eggs in one basket." Suddenly it seems as if the number of "baskets" has been substantially reduced.

All of us in this industry have been trained to avoid creating single points of failure. That's why we spread our risk and hedge our bets. I'm delighted that big U.S. companies like Apple, Amazon, Google, and Netflix are having great years and generating incredible value for their shareholders. Like most of us in the technology industry, I'm in favor of strong competition at every level. Competition drives innovation and innovation is essential for a healthy economy.

Apple's ascent overshadowed a report in the Wall Street Journal that GE is now planning to sell its digital assets.[3] For those of us who had been following GE's entry into the software business, this is disappointing news. Under the leadership of former CEO Jeff Immelt, GE invested billions to create and grow a digital business unit. It also developed Predix, a specialized digital platform for the Internet of Things. Clearly, GE's new leadership has a different vision of the future.

The plan to reposition GE as a global software company didn't work out. A lot of talented people poured their hearts and minds into the effort, but success eluded them. Despite its investments and good intentions, GE couldn't keep pace with the speed of change in the 21st-century software industry.

If there's a moral to the story, it's that success in today's markets really does depend on speed and execution. It's not enough to have a great idea; you need to bring your idea to market before someone else beats you to the punch.

Apple understands why speed and execution are absolutely imperative. That fundamental understanding is deeply woven into Apple's corporate and cultural DNA. Amazon, Google, and Netflix have similar cultures, which is why their valuations have also reached stratospheric heights.

How Will the "Amazon Effect" Impact Your Company?

In case there was still any doubt, Amazon has pulled away from the pack and is now in a league of its own. The term "Amazon Effect," which was invented to describe Amazon's transformative impact on retailing, now applies to virtually every aspect of business competition.

From my perspective, the key to Amazon's continually amazing performance is discipline. With Jeff Bezos at the helm, Amazon has remained steadfastly focused on providing its customers with experiences that are consistently superior to any of its competitors. Amazon knows how to keep its customers happy. Even after testifying to Congress, Mark Zuckerberg still seems clueless about the absolutely critical role of customer experience in modern business.

If you're not providing the best possible customer experiences across your enterprise, you aren't in the game. Amazon understands this. Facebook doesn't.

As technology leaders, we need to emulate the discipline shown by Amazon and keep our attention focused on providing the best possible customer experiences. Anything else is really a distraction.

Recent market gyrations confirmed Amazon's growing dominance as a primal force driving the modern global economy. The Amazon Effect has spread from retail to every imaginable part of the competitive landscape. One of the reasons Rupert Murdoch is selling off chunks of his media empire is because he understands that he no longer has the resources to compete effectively with companies like Amazon and Netflix.

I predict that Amazon and Netflix will continue growing their clout as creators of original content. Within years, Amazon and Netflix will be more like Disney—diversified ecosystems of content and entertainment offerings spanning an incredibly wide range of media platforms. AT&T's acquisition of Time Warner will seem pale in comparison to what happens next.

We're at an inflection point in history. The old giants have been replaced by new giants, and now the new giants are battling for leadership. The stakes are high, and the game is still in its early stages.

The unknown factor in this new landscape is China. What role will China play? It's hard to imagine that China will sit by idly while companies like Amazon and Netflix carve up the market. China has the scale and the energy to compete, but do its leaders have the vision?

Again, we're early in the game and it's too soon to tell. But my prediction is that Amazon and Netflix will discover their main competition is China. Google learned that lesson several years ago and has all but given up competing directly in China.

We live in interesting times, and the real competition is just beginning to heat up. Are you focused on the customer experience? Is your enterprise ready to compete against Amazon?

Rise of AWS Influences the Broader Economy

With a booming economy and nearly full employment, recent holiday shopping seasons have brought joy to many retailers. But most of those retailers are also looking over their shoulder at Amazon, which continues to expand its dominance in consumer and industrial markets.

Vandana Radhakrishnan, a partner in Bain & Company's New York office, described Amazon's focus on holiday sales as "truly maniacal" during a CNBC interview.[4] Amazon's strengths, she said, include speed, price, and exclusivity through carefully chosen private label brands.

I firmly believe that Amazon's rise as an e-retailer is part of a much larger and significantly more complex story. That's the reason why technology executives need to focus on Amazon and learn from its example.

For instance, Amazon doesn't limit itself or set artificial boundaries on its own growth. From Amazon's point of view, the entire world economy is its playing field. Amazon will compete for market share wherever it perceives an opportunity.

Who would have predicted that a website that was initially devoted to selling books would become the world's largest provider of cloud services through Amazon Web Services (AWS)?

As technology executives, we need to zoom in and genuinely understand the consequences of Amazon's rise as a tech vendor. On one hand, it's great that Amazon makes it possible for companies large and small to use its genuinely vast computational resources. On the other hand, does it make sense to expose your data to an enterprise with truly unbounded ambitions?

There's a growing sense that partnering with Amazon could backfire, especially for companies in potentially volatile industries, such as health care and retail grocery, which are ripe for disruption.

"AWS is the world's biggest public cloud … But as Amazon expands into countless new areas, from grocery stores to health care, some companies that have previously worked with Amazon have found a partner becoming the competition overnight," wrote Jordan Novet of CNBC.[5]

Novet's article is worth reading closely, since it also reveals the extent to which AWS impacts the development of open-source software. Open source has been the "secret sauce" of innovation in the information technology business for more than 20 years, even though most people outside the industry are unaware of its critical importance within the global IT ecosystem.

Amazon's dominance will have multiple consequences across numerous sectors of the economy. No industry is safe from disruption. If Amazon believes it can compete successfully against you, it probably will.

Apple Takes on Role of World's Privacy Watchdog

Technology drives the world's economy. That is a simple fact of modern life. But technology is also driving vast changes in culture, all over the world. Inevitably, the lines between technology and culture are blurring.

Here's a case in point: In January 2019, Apple revoked and then restored enterprise developer certificates for both Facebook and Google. According to Apple, both tech titans had violated Apple policies designed to protect data privacy.

The missteps were quickly corrected, and by the end of the week, the developer certificates had been reissued, apparently ending the dispute.

But a much larger question remains unresolved. Are we comfortable with the idea of Apple serving as the world's privacy watchdog? Personally, I applaud Apple's efforts to keep my data safe and secure. But I wonder if delegating the role of watchdog to Apple is really the optimal way for protecting our data privacy.

As leaders in the technology community, I believe we need to become more involved in the debate over who is—and who isn't—responsible for data privacy.

People who say there is no such thing as privacy any more are missing the point. In a world in which data is the new oil, personal data has real economic value. When organizations misappropriate or misuse our personal data, trust erodes and the fabric of society becomes weaker.

Eventually, governments will catch up with technology and create better rules for governing the collection, analysis, and sharing of personal data. Until then, we're living on the edge of a new frontier. I'm glad that Apple has taken some of the responsibility for safeguarding our data, but it would be unwise to delegate the job to Apple on a permanent basis. We need a long-term solution to the challenge of data privacy. As tech leaders, we need to raise our voices and join the debate. Data privacy is our collective responsibility, and we cannot delegate it to a single company or organization.

China Rising: Is the United States Losing the Battle for AI Dominance?

Today we're on the cusp of a new revolution that will radically transform information technology *and* human culture itself. When historians of the future look back at the early 21st century, they will wonder if we realized how profoundly our world would change.

For senior technology executives, the rise of artificial intelligence offers multiple challenges and opportunities. I guarantee the CEO, CFO, COO, CMO, and board of directors will be asking you for updates and advice on the latest developments in AI.

Additionally, you will be fielding intense questions from your team, many of whom will be concerned over the potential for AI to eliminate their jobs in the near future.

Make no mistake: AI will rock the foundations of our existence. Soon most of the tasks we consider part of our daily lives will be performed by smart machines.

Driving, shopping, cooking, cleaning, paying taxes, and registering your children for kindergarten will be fully automated. Instead of going to your doctor's office for a checkup, you'll merely place your fingertip on a special touchpad on your computer at home and instantly receive a diagnosis, as well as prescriptions for any medications you might need.

The heart and brain of this revolution is AI, which arose from years of dormancy to become one of the most important economic forces on the planet.

But here's the unexpected twist: The AI revolution isn't evenly distributed. Almost all of the world's AI innovation and development is taking place in two countries: the United States and China.

I recently watched a fabulous interview on CNBC with Kai-Fu Lee, chairman and CEO of Sinovation Ventures, a leading investment firm focused on developing the next generation of Chinese technology companies.[6] In the interview, Lee warned of a looming age of inequality as China and the United States compete for dominance in the AI realm. AI is already becoming an "inequality multiplier," Lee said, citing the shortage of affordable housing in San Francisco as an early example.

Lee isn't worried about the ultimate economic fates of China and the United States. The rest of the world, however, will find it extremely difficult to compete with the entrenched

AI "superpowers." That's where the inequality will become problematic, he says.

As you would expect, Lee is bullish about China's prospects as an AI leader. After listening to the CNBC interview, I ordered a copy of his new book, *AI Superpowers: China, Silicon Valley, and the New World Order.*[7] The book is definitely worth reading, and I recommend it highly.

"When Chinese investors, entrepreneurs, and government officials all focus in on one industry, they can truly shake the world," Lee writes in his book. "Indeed, China is ramping up AI investment, research, and entrepreneurship on a historic scale. Money for AI startups is pouring in from venture capitalists, tech juggernauts, and the Chinese government."

All of China, he suggests, is fascinated with AI. "Chinese students have caught AI fever as well, enrolling in advanced degree programs and streaming lectures from international researchers on their smartphones. Startup founders are furiously pivoting, reengineering, or simply rebranding their companies to catch the AI wave," he writes.

Lee also makes an important point around the different waves of AI. He says major shifts won't all manifest at once. They will instead come in waves: internet AI, business AI, perception AI, and autonomous AI. The first wave of internet AI appears around us today in the form of recommendation engines. The second wave of business AI mines vast amounts of business data to find hidden correlations. Perception AI enables algorithms to recognize objects in pictures and videos.

Last, the fourth wave of autonomous AI brings together the three previous waves. There aren't many companies able to deliver on this fourth wave of AI, because of the complexity of solutions required, but they do exist and are already making an impact at many enterprises.

Notes

1. Kevin Hartnett, "Finally, a Problem That Only Quantum Computers Will Ever Be Able to Solve," *Quanta*, June 21, 2018, https://www.quantamagazine.org/finally-a-problem-that-only-quantum-computers-will-ever-be-able-to-solve-20180621/

2. Ben Rossi, "Digital Disruption Will Wipe Out 40% of Fortune 500 Firms in Next 10 Years, Say C-Suite Execs," *InformationAge*, February 17, 2017, https://www.information-age.com/65-c-suite-execs-believe-four-ten-fortune-500-firms-wont-exist-10-years-123464546/

3. Dana Cimilluca, Dana Mattioli, and Thomas Gryta, "GE Puts Digital Assets on the Block," *Wall Street Journal*, July 30, 2018, https://www.wsj.com/articles/ge-puts-digital-assets-on-the-block-1532972822

4. CNBC, "Private Labels, Fast Shipping Drives Amazon's Dominance, Says Bain Partner," *Squawk Box*, November 30, 2018, https://www.cnbc.com/video/2018/

11/30/bain-and-co-partner-on-what-drives-amazons-dominance-in-retail.html

5. Jordan Novet, "Amazon's Cloud Business Is Competing with Its Customers," CNBC, November 30, 2018, https://www.cnbc.com/2018/11/30/aws-is-competing-with-its-customers.html

6. CNBC, "The US and China Race for Artificial Intelligence Leadership," *Squawk Box*, February 4, 2019, https://www.cnbc.com/video/2019/02/04/the-us-and-china-race-for-artificial-intelligence-leadership.html

7. Kai-Fu Lee, *AI Superpowers: China, Silicon Valley, and the New World Order* (New York: Houghton Mifflin Harcourt, 2018).

Chapter 9

Key Takeaways

We learned many extremely valuable lessons as we conducted our research for this book, the sixth in a series of leadership books conceived and created by the team at HMG Strategy LLC. Here are some of the main points and key takeaways from our comprehensive and truly unique method of peer-to-peer research:

Growing the Business

- CIOs and CDOs can do more than support new business initiatives. They can also act as catalysts for identifying and executing on new business opportunities.

- In order for a company to disrupt itself, it has to shift its mind-set completely and be willing to try bold and daring new ideas—to think and act differently.

- While a CIO or another CXO can lead business disruption, creating a cross-functional team to brainstorm on ideas and provide different perspectives is critical for enterprise-wide buy-in and success.

Digital Transformation

- Going digital offers executives an opportunity to reimagine and reinvent the business, in terms of both crafting new business models and creating operational efficiencies via automation.

- Machine learning and other forms of automation offer incredible opportunities to automate standard repeatable tasks, giving back time to focus on higher-value activities.

- CIOs should shift away from sharing look-back responsive-based metrics such as speed of service response. Instead, they should focus on demonstrating to the CEO measurable improvements in efficiency and net promoter scores.

Leveraging the Power of Customer Data

- Customer journey mapping, including the use of customer behavioral data and sentiment analysis, can provide deep insights into customer preferences and needs that project teams can act on to provide customers with richer experiences.

- Delivering on customer expectations also requires executives to understand customer-facing processes along with any challenges employees may face in their roles that may create friction for customers.

- It's extremely valuable for CIOs and members of their IT teams to spend time on the frontlines of the business to better understand and respond to employee and customer needs.

Focus on the Customer Experience

- As much as company executives have learned about their customers through a variety of customer data that's available, there are always opportunities to go wider and deeper.

- Every company in every industry is a technology company. Step up as a CIO and play a leading role on the executive team in setting the company's strategic course.

- Empower members of the IT team to collaborate with their business peers on innovation. Provide them with the tools, the time, and your support in helping them to succeed.

Diversity and Innovation

- Female IT leaders can and should encourage young women to explore IT and STEM professions, in part by helping to clarify any preconceived notions young women might have about working in IT and by pointing to examples of female role models.

- To help encourage female IT professionals to remain in their roles, female IT leaders can help teach them how to advocate for themselves, including steps they can take to be considered for interesting assignments and promotions and with regards to work–life balance.

- To succeed as technology professionals and accelerate their career ascent, women—as well as men—need to be willing to go outside their comfort zones.

Next-Generation Technologies

- Test new technologies and launch small-scale pilot projects to determine if they are appropriate for your organization.

- Create a program of seminars, lectures, symposia, and presentations on new technologies and techniques to maintain an atmosphere of continual learning and innovation.

- Develop an in-house awards program to recognize experimentation that doesn't necessary result in economic benefits but that increases the knowledge and capabilities of the organization.

- Sponsor hackathons, coding festivals, and workshops to encourage innovation, unleash potential, and normalize risk-taking.

- Whenever possible, gamify internal work processes to inspire greater engagement and focus on excellence.

Communicating ROI to the C-Suite

- CIOs in industries that are undergoing dramatic disruption are well-positioned to communicate to the CEO and the board those technologies that can help move the business forward.

- Clearly communicate the ROI achieved through technology-enabled business initiatives to demonstrate the value that the IT organization consistently brings to the table.

- Working with a trusted partner can help you and your organization to identify cybersecurity vulnerabilities and issues you may not even be aware of.

Inspirational Leadership

- As an organizational leader, part of the role of a CIO/technology executive is to move employees out of their comfort zones and to inspire and motivate them to take on new challenges.

- There's no "I" in *team*. The success of the business is dependent on the ability to function and execute together as a team.

- As a member of the C-suite, come armed with approaches to innovate the business and to create new business models.

- CIOs, CDOs, and IT teams must immerse themselves in the customer experience to understand and respond to customer needs and motivations.

Explaining the Need for Disruption

- It's healthy to be paranoid about disruption—disruptive forces can attack your company from within your industry or from external players.

- If the current strategy isn't working, have the courage to speak up and instigate change.

- Board members, like other people, respond to visual stimuli. Conducting a tech fair to demonstrate the business value that technology investments can deliver is a great way for CIOs and their teams to get their messages across.

Leverage Social Media and Professional Networks

- Use social media to raise awareness of your organization's successes and to praise the accomplishments of your colleagues.

- Share relevant articles about new developments in technology.

- Make it a habit to post at least once a week on LinkedIn and other professional networks.

- Write short articles on technology leadership topics and post them to LinkedIn.

- Create a website for your organization, and encourage staff to make regular contributions.

- Produce short videos about your team's accomplishments, and post them on YouTube or Vimeo.

- Make yourself available for speaking engagements and professional presentations.

- Attend professional conferences, such as HMG Strategy CIO/CISO Leadership Summits.

ABOUT THE SOURCES

Eric Anderson is the managing partner and global head of Tech Enabled Services Practice for the Egon Zehnder Atlanta office and the head of the FinTech practice group. He is a core member of the Financial Services and Technology Practices serving industry clients in executive search, board search, and management appraisal capacities. He is also a member of the Financial Officers practice having accomplished numerous senior-level assignments. Prior to joining Egon Zehnder International, he was the president of BISYS Compliance and Licensing Services, where his responsibilities included growing the Education, Licensing and Advanced Designations' business for BISYS. BISYS was sold to Citibank in 2005. Eric has a BS in Marketing from the Pennsylvania State University and an MBA in International Business from the University of Miami.

Snehal Antani is an entrepreneur, technologist, and investor. He is chief executive officer and co-founder of Horizon3.ai, a cybersecurity company using AI to deliver red teaming and penetration testing as a service. He also serves as a Highly Qualified Expert (HQE) for the US Department of Defense, driving digital transformation and data initiatives in support of Special Operations. Prior to his current roles,

Antani was chief technology officer and senior vice president at Splunk, held multiple CIO roles at GE Capital, and started his career as a software engineer at IBM. He has a Masters in Computer Science from Rensselaer Polytechnic University, a BS in Computer Science from Purdue University, and holds 16 patents.

Renee Arrington is president and chief operating officer at Pearson Partners International. She is a strategic thinker who connects the dots and presents executive talent for Fortune 500 companies, private equity–backed businesses, and not-for-profit organizations. She works across many industries including technology, retail, financial services, business services, manufacturing, distribution, and R&D. Renee serves as Vice Chair Americas for Pearson Partners International global search organization, IIC Partners. With 43 offices in 29 countries, IIC Partners is one of the world's top 10 executive search organizations. Additionally, she serves on the board of directors of the National Association of Corporate Directors North Texas Chapter and the Foundation for Young Women's Leadership Academy in Fort Worth, which encourages young girls to excel in STEM. She is a member of Dallas Executive Women's Roundtable and past member of Fort Worth's Women Steering Business. Renee earned a B.A. degree in communications from Trinity University and studied at Parsons School Design in Paris.

Shankar Arumugavelu is the senior vice president and global chief information officer of Verizon Communications, a leading provider of wireless, fiber-optic, and global IP network services. He is responsible for the company's information technology strategy, architecture, development and management of the information systems portfolio, continued evolution of digital platforms, and operation of all supporting infrastructure.

Before being named to his current position in 2017, he served as senior vice president and chief information officer for Verizon Wireless and Verizon Consumer Markets. Previously, he was the senior vice president and chief information officer for Verizon's wireless business unit. Prior to this, he has held a number of positions of increasing responsibility at Verizon and its predecessor companies. Immediately before his chief information officer assignment, he was senior vice president, Network Operations Support Systems for Verizon Telecom. He was responsible for systems strategy, architecture, design, development, implementation and maintenance of engineering, service provisioning, service assurance, workforce management, dispatch, testing, surveillance, and performance management systems for voice, data, video, special access, and new generation communication services.

He began his career at GTE Data Services in 1997 as a member of technical staff in the Enterprise Systems Group and advanced to increasing levels of responsibility in IT. In 1998 he was named chief architect of the group and was responsible for the design, development, and maintenance of GTE's service provisioning system. He was appointed as the executive director of the Provisioning and Activation Systems group in 2002.

Arumugavelu was recognized as one of the "Premier 100 IT Leaders" by *Computerworld* in 2014. He earned a bachelor's degree in Electrical Engineering from Anna University (India) and a master's degree in Computer Science from the University of South Florida.

Ashwin Ballal, Ph.D., is senior vice president and chief information officer at Medallia, a market leader in customer experience management in 2016, and its first-ever CIO. He is responsible for building best-in-class IT infrastructure, collaboration services, workplace services, and business systems that globally scale and provide a spectacular employee and digital systems experience. During his tenure at Medallia, he has been responsible for Customer Experience Management, including several transformational business programs and projects that drive the top line growth and go-to-market strategies. He also envisioned and developed a unique product solution for CxOs. Prior to joining Medallia, Ashwin was chief information, intelligence, and data officer at KLA-Tencor, a market leading semiconductor capital equipment company globally responsible for driving business value; managing and supporting IT Infrastructure, business applications, productivity, and collaboration tools; advanced data analytics; and enterprise risk management. He earned his doctorate and postdoctorate degrees in material science and engineering from the University of Maryland at College Park and his bachelor's degree in metallurgy from the National Institute of Technology in India.

Shawn Banerji is the managing partner for the Technology, Digital and Data Leaders Practice at Caldwell Partners. As a trusted advisor to organizations seeking to transform their businesses by unlocking the latent value of their legacy product portfolios, he advises clients on the convergence of talent and innovation across all facets of the evolving digital enterprise. He specializes in the recruitment and assessment of these catalyst leaders who are resetting the value propositions of their organizations and industries through the power of data and analytics, accelerated software engineering, and cloud. Shawn is a loud advocate for increasing technology representation and diversity on corporate boards. He is also an ex-officio board member of SIM New York and serves on the board of advisors for the Center for Technology Management at Columbia University.

JP Batra is chief technology officer and chief information officer of Product and Program Management at Blue River International. He is a technology executive and a thought leader who has revitalized IT departments, turned around stalled strategic initiatives, and managed them to success. His work has resulted in business acceleration using technology, better employee engagement, and improved time to market. His successes include reenergizing revenue, stock, and reputation impacting initiative for a Fortune 500 client and applying emerging tech to gain market share over entrenched services leaders for his client company.

David Bessen is director and chief information officer of Arapahoe County, CO. There he has established IT governance, converted the IT Department to a bimodal operation, and focused on using innovative technologies to drive government efficiencies and create collaboration. Recently his department's work was awarded a CIO Digital Edge 50 Award, in recognition of the county's game-changing use of technology that benefits citizens in both Arapahoe and other Colorado counties. Prior to joining the county, he was in the media (newspaper and website) industry, where he served as vice president and chief information officer at MediaNews Group, the second largest newspaper chain in the United States (headquartered in Denver), and IT director at Copley Newspapers in La Jolla, CA.

Steven Booth is vice president and chief security officer at FireEye, where he is responsible for leading all aspects of IT security. This includes the ongoing development and execution of enterprise-wide security architecture and monitoring programs as well as technology risk and compliance. Steven directs the implementation of security controls, standards, policies, and procedures to ensure continuous monitoring and protection of information systems and physical property while ensuring that compliance to both corporate security policies and industry standards is maintained. Prior to joining FireEye, Steven was the assistant vice president of Information Security and CISO at Manulife/John Hancock in Boston, MA, where he had global responsibility for all IT security in the United States, Canada, and 11 countries in Asia. He has a B.S. in business management from CSUH and a master's degree in digital security from Golden Gate University.

Chris Colla, vice president and CIO of B&G Foods, Inc., is an active member of the New Jersey chapter of the Society of Information Management. He is a respected technology and process management executive with 20 years' experience leading organizational change and operational excellence through innovative technology solutions. He has held IT leadership positions at Accenture, Diageo, Sharp Electronics, and Haier America before B&G Foods. Mr. Colla has a bachelor of science in mathematics from the University of Mount Union in Alliance, OH, and a master of science in information management from the Howe School of Business at Stevens Institute of Technology. He also has graduate certifications in project management and business process management from Stevens Institute.

Bob Concannon is a senior client partner within the San Francisco office of Korn Ferry. He is a member of the firm's Global Technology Market. Mr. Concannon brings to Korn Ferry over 10 years of technology executive search experience, leading numerous CIO and CTO searches across industries. He also serves on Korn Ferry's Information Technology Officers Center of Expertise. Furthermore, Bob leads searches in the software and IT communities on assignments for clients in the areas of cloud, mobility, and enterprise software. Venture capitalists and private equity firms look to his expertise in helping internet startups migrate to the next executive level as they receive their post-Scries B and C funding rounds. Prior to his current position, Bob served as the managing director of another global executive search firm, leading the technology practice. In addition, he led the firm's global technology vertical for seven years, coordinating worldwide search activity in the technology sector with Fortune 100 top accounts. He has a bachelor's degree from the University of the Pacific.

Barbara Cooper is the former CIO (retired) of Toyota Motor North America. She is an innovator and risk taker, known for her passion for business alignment, organizational strategy, and optimizing and developing exceptional IT talent. She became the first female vice president in the technology function at American Express and led the company globally into the personal computing and networking generation, connecting travel offices and operating centers around the world for the first time. Her reputation as an innovator and risk taker built from there as she took on her first CIO role at Maricopa County, AZ, the fourth largest county in the United States.

Once her reputation as a successful turnaround and transformational leader was established, Barbara was recruited for even more challenging CIO roles. After a brief time as the CIO for MicroAge, she was recruited by Toyota to transform an IT function that was not keeping pace with the dramatic growth of the business in North America. As group VP and CIO of Toyota, she was responsible for modernizing the company's technology and governance practices. She established her own business in 2013, specializing in executive coaching for CIOs and senior IT management, in addition to consulting for IT organizations in transition. She serves as a coach for the CIO Executive Council's Pathways Program and coaches clients directly in both private and public sector IT as part of her consulting company.

Jamie Cutler is senior vice president and chief information officer at Air Methods. He is a 17-year veteran in technology and corporate management and over his career has held leadership positions in the public, hospitality, technology, and most recently the oil and gas sectors. Mr. Cutler previously served as the chief information officer at QEP Resources Inc., an S&P 500 energy company based in Denver, CO. Before that he was with MarkWest Energy Partners in Denver.

Dale Danilewitz is a self-employed senior executive who most recently served as executive vice president and chief information officer at AmerisourceBergen, a Fortune 11 health-care company. He originally joined the company in 1999 as vice president of Information Technology. Prior to joining AmerisourceBergen, he held management positions with American Airlines and the Sabre Group and spent three years working for Whirlpool Corporation in the Advanced Technology Group. Dale holds undergraduate and graduate degrees in computer science from South Africa and a master of science in electrical engineering from the Imperial College of London.

Daniel Dines is cofounder and CEO of UiPath, the fastest-growing and leading provider of robotic process automation and AI software worldwide. He started UiPath in 2005 with the goal of building a solution that could help humans reduce the time and stress that come from menial, administrative business tasks. Today his vision is to make software robots, powered by computer vision and AI, as common as PCs in the workplace, a vision dubbed "One Robot for Every Person." Daniel began his engineering career at Microsoft, where he designed and developed the SQL server agent. Still a hacker at heart, he is active in driving the company's RPA platform and supporting the 250,000 developers who are building automations on UiPath's RPA platform.

Amy Doherty serves as executive vice president and chief operating officer at First Command Financial Services, Inc. Previously, she was senior vice president and chief information officer at AARP. Amy started with AARP in December 2011, providing leadership to application portfolio management and enterprise resource planning. She has worked in roles of increasing responsibility in information technology services. She has an expansive view of diversity and seeks to build teams that include members with diversity of thought and diversity of experience. She has a passion for STEM and works to encourage women to join the IT field and stick with it. Amy actively advocates for and speaks about how women in tech can be better supported. She also proudly serves as a Sequoyah Fellow with the American Indian Science & Engineering Society.

Nicole Eagan is chief strategy officer and AI officer of Darktrace. Her extensive career in technology spans 30 years working for Oracle and early to late-stage growth companies. Nicole identifies and shapes Darktrace's strategic plan, leads the Company's AI vision together with our CTO, and provides product strategy and direction. A core part of the executive team, during Nicole's tenure, Darktrace has won more than 100 awards, and the company has been named one of WSJ's "Tech Companies to Watch", Fast Company's "Most Innovative Companies", and a CNBC "Disruptor 50". Nicole was named "AI Leader of the Year" and was awarded the top position on The Software Report's "Top 25 Women Leaders in Cybersecurity" in 2020.

Mark Egan is a partner at StrataFusion Group, Inc. and has more than 25 years of information technology experience, with expertise in IT transformation, information security, and mergers and acquisitions. His extensive experience includes managing global IT organizations with over 1,200 employees, budgets in excess of $350 million, and IT integration for over 60 mergers and acquisitions. He has been repeatedly recognized for his leadership and success in scaling IT organizations to support rapid growth companies. Prior to joining StrataFusion, Mark was a CIO at VMware, where he led the company from a server virtualization vendor with $2 billion in revenue to a $5 billion market leader of cloud solutions. He is author of *Executive Guide to Information Security: Threats, Challenges, and Solutions* and was a contributing author to *CIO Wisdom, CIO Perspectives,* and *CIO's Body of Knowledge.* He is the founder of the CIO Development Program, which mentors aspiring CIOs, and president of the CISE Education Fund, which raises money for underprivileged students to pursue technology degrees. He holds a master's degree in finance and international business from University of San Diego and a bachelor's degree in computer science from Clarion University.

Thomas Farrington is executive vice president and chief information officer of Perrigo Company plc. He also oversees Perrigo's Corporate Social Responsibility program, Global M&A Integration, and its e-commerce business unit. Thomas has led the company's integration efforts over 27 acquisitions resulting in expansion into more than 35 countries; built out a global shared services platform; launched the company's direct-to-consumer sales, breaking into new markets and innovative retail opportunities; and has gained end consumer insights relative to over-the-counter health-care purchases. Prior to Perrigo, Thomas served as chief information officer for the F. Dohmen Co. and served as a division president for JASCORP LLC.

John Fidler is managing director of Retained Search at Fidato Partners, LLC. He leads Fidato's Retained Search Practice focused on placing executive-level information technology, finance, and accounting professionals in the Mid-Atlantic region. Prior to joining Fidato, he spent time with a major global search firm, where he worked with a number of private equity–backed, publicly traded, and Fortune 1000 clients. Previously John was a partner at a boutique retained search firm, where for more than eight years he focused on client service and business development and led the successful placement of IT and finance executives for a wide variety of companies across all industries. John began his career overseeing the sales and marketing function for a wholly owned subsidiary of Random House. He spent time living and working overseas helping to build the company's offshore operation. During this time, John recruited and built a majority of the company's salesforce. John has presented to a number of local outplacement and industry associations and has published several articles on topics related to the recruitment of senior executives. He is deeply entrenched in the Philadelphia business community and has worked closely with organizations such as the Walnut Street Theatre and Philadelphia Outward Bound School. He is a graduate of Bloomsburg University.

Hugo Fueglein is a managing director in Diversified Search's New York City office. He is a leader in the CIO and IT Management Practice in the United States and is a core member of the Global CIO Practice for Diversified Search. He is also a leading member of the Technology and Innovation Practice at Diversified Search, having placed technology executives in key positions across all industries. His specialization is in the placement of chief information officers, chief information security officers, chief technology officers, and technology executives in IT architecture, IT infrastructure, application development, and IT operations. Prior to joining Diversified Search, Hugo was an executive vice president with DHR International and an integral member of the CIO and IT Practice at CTPartners. He served clients across industries, recruiting talent to fill critical roles in corporate technology functions. His search career also includes being a principal with both Korn Ferry International and Morgan Samuels. During his tenure with Korn Ferry, he was instrumental in the creation of the CIO Center of Expertise.

Curtis Generous serves as chief technology officer and vice president of Engineering at Earth Networks. As CTO, he has responsibility for the overall leadership of Earth Networks' enterprise technology initiatives, including software and hardware development and information technology. Prior to Earth Networks, he was CTO at AARP, where he had responsibility for IT operations, software development, data and network communications, enterprise architecture, and information security. In that capacity, Curtis drove technology modernization initiatives and migration to cloud/SaaS services, with a heavy focus on innovation strategy. Prior to his role at AARP, he served as CTO of Navy Federal Credit Union, leading the advanced technology and enterprise architecture groups. Earlier in his career, Mr. Generous was a hardware design engineer and founded his own technology firm after a successful research engagement at Lawrence Livermore National Labs. He is a graduate of Cal State University with a B.S. in Industrial Technology. Also, he did advanced degree work in electrical engineering at Loyola and is a retired Naval Reserve Officer.

Marc Hamer is a talented and experienced chief information officer and chief digital officer currently working in private equity. He served as corporate vice president, global chief information officer, and chief digital officer for Sealed Air, an innovative packaging company known for its Cryovac food packaging and Bubble Wrap cushioning packaging brands. In addition to fueling innovation, Marc was responsible for Sealed Air's IT systems and assets globally, ensuring that technologies align with the objectives and priorities of the business. He was also responsible for the process improvement, automation, and digital initiatives throughout the company to streamline operations and improve bottom-line profitability. He was named one of the 2019 HMG Strategy Top Technology Executives to Watch. He sits on advisory boards for the Global CIO Executive Summit, PTC, Insight Venture Management, FireMon, and Nutanix. He is also a member of the CNBC Technology Executive Council.

Jon Harding is the global CIO of Conair Corporation, a multibillion-dollar consumer product company focused on personal care products, countertop kitchen appliances, and health and beauty products used by professionals and consumers. He has worldwide responsibility for the company's IT and digital services in support of both day-to-day business operations and future business growth. During his tenure, the Conair IT team has completed global rollout of a centralized ERP system replacing 25 different legacy ERP systems with one single platform (SAP); global business integration via a standardized IT network; seamless integration of eight acquired businesses; and supported opening of seven new businesses. The company has doubled revenue and increased global reach multiple times during this period. Before joining Conair, Jon was divisional CIO for the U.S. Snacks Division of Kellogg Company (formerly the Keebler Company, which Kellogg's acquired in 2001).

A graduate of the University of Sheffield in England, he has 25 years of IT management experience in the consumer products industry. He shares this experience industry-wide through his membership of the Executive Advisory Board of Consumer Goods Technology and industry-leading e-commerce and CIO strategy events. Mr. Harding cofounded and chairs the Princeton CIO Roundtable that meets quarterly for peer group sharing of ideas and experiences across industries and all digital technologies.

Vishwa Hassan is director of Data and Analytics at USAA. He has over 20 years of experience, having held various technical and managerial positions in leading enterprises, most recently at Intel. He is engaged in accelerating USAA's competitive advantage through data and platform transformation in partnership with business, IT, and external organizations, including solution integration partners. Prior to joining Intel, Vishwa cofounded a startup focused on providing managed business intelligence services to mid-market companies. He has over 12 issued patents to his credit. He received his master's degree specializing in reliability engineering from the Université d'Ottawa, Canada.

Patty Hatter is senior vice president of Global Customer Services at Palo Alto Networks, where she is driving the transformation to enable new SaaS business models. She also serves on the board of directors for II-VI (NASDAQ:IIVI). She is a multiple award-winning technology, business, and operations leader who drives digital transformation and growth for global companies. Prior to Palo Alto Networks, she was SVP of the McAfee Services organization, where she drove 50 percent growth. At McAfee Services, she also served as general manager and CIO for Intel Security as well as CIO and SVP of Operations. There she orchestrated a global transformation of IT and operations. Recognized by Gartner as leading the transformation of an underperforming IT organization to one that beat all Gartner's metrics for peer companies Patty delivered 15 percent reduction in IT spend vs. revenue while simultaneously enabling increased spend/delivery for new capabilities by 25 percent. In recent years Patty has received a series of highly coveted awards, including "Women of Influence," "Power Executive," and "Bay Area CIO of the Year" by *Silicon Valley Business Journal*.

Jason Hengels, CISSP, CISM, CISA, CRISC, founded Exposure Security in 2014 in response to growing requests from companies around the San Francisco Bay Area for veteran security leadership. Throughout his career, he has led companies through technical debt reduction, compliance, incident response, and general remediation efforts. He is a pragmatic security leader who comes from a strong technical security background. As VP of Security at Box, he grew the security program from a one-man show into a team of industry experts in under two years. He also held executive security leadership roles at Visa, CyberSource, and Authorize.net. Jason feels compelled to do everything he can to help the industry progress. Over the last two years, Jason has also worked with top-tier cybersecurity veterans to create an information security program at Merritt College in Oakland, CA. He is an adjunct professor at Merritt College, where he teaches all of the application security courses.

Donagh Herlihy is executive vice president and chief technology officer for Bloomin' Brands. In this position, he is responsible for overseeing the global strategy, development, and implementation of all technology initiatives for Bloomin' Brands. He has 25 years of technology leadership experience in various industries. Donagh joined Bloomin' Brands following a six-year tenure with Avon Products, Inc., where he served as CIO and SVP of e-commerce. In that role, he led all aspects of the technology supporting Avon's internal business operations and enabled e-commerce for more than 6.5 million Avon representatives worldwide. Donagh has a BSC and MA in industrial engineering from the Institute of Technology and Trinity College in Dublin and has completed the Executive Program at the University of Michigan, Ross School of Business. In 2016 he was inducted into *CIO* magazine's CIO Hall of Fame.

Sheila Jordan is senior vice president and chief information officer at Symantec. She is responsible for driving Symantec's information technology strategy and operations, ensuring that the company has the right talent, stays ahead of technology trends, and maximizes the value of technology investments. Her goal is to drive increased productivity, better efficiency, and strategic business partnerships through simple and intuitive experiences for Symantec's global workforce. Prior to joining Symantec in February 2014, she served as senior vice president of Communication and Collaboration IT at Cisco Systems. There she was responsible for delivering and integrating key IT services that touched Cisco's global workforce. During her nine-year tenure at Cisco, Ms. Jordan's IT organization was recognized by the industry with awards presented by CITE, *InformationWeek*, and *CIO Magazine*. She also received the 2010 Cisco Executive Sponsor Catalyst award for her leadership in developing successful relationships with customers, including several Fortune 100 companies. Active in social media, this year Sheila was recognized by ZDNet as one of the Top 100 CIOs on Twitter. In 2011 she was named by *CIO magazine* as "One to Watch." She serves on the board of directors for NextSpace, a Santa Cruz, CA–based firm that provides innovative physical and virtual infrastructure for entrepreneurs, freelancers, and creative professionals. She holds a B.A. degree in accounting from the University of Central Florida and an MBA from Florida Institute of Technology.

Chad Kalmes, CISA, CISSP-ISSMP, is vice president of Technical Operations at PaperDuty, the leading SaaS platform for real-time operations and incident response. He is a proven executive with over 20 years of technology, security, and Big 5 consulting experience. Throughout his career he has worked with companies large and small—from members of the Fortune 500 to pre-IPO technology startups—to strategize, implement, and refactor their operations for hypergrowth, public company readiness, and long-term scale. He has built successful teams spanning production operations/engineering, security and compliance, business intelligence, and IT. Chad holds a BBA in management information systems from the University of Notre Dame.

Tom Keiser has been chief operating officer for Zendesk since August 2017. Tom served as the company's chief information officer from May 2016 until his appointment as COO and as the company's senior vice president of Technology Operations from October 2016. From January 2014 to March 2015, Tom served as executive vice president of Global Product Operations at The Gap, Inc., where he had been chief information officer. Tom holds a B.S. in systems science from the University of West Florida.

Steven Kendrick is president of KER Partners, LLC. His background includes over 25 years of successful retained executive search industry and corporate recruitment professional experience. This experience includes leading successful search engagements and client relationships across commercial industry segments, including financial services, banking, payment systems, consumer retail, online e-commerce, logistics-transportation, industrial, pharmaceutical, and health care. He has served industry-leading clients, including global

Fortune 500 multinationals, mid-cap-mid-market public/ private firms, and private equity–backed portfolio companies. Prior to KER Partners, he was a partner at the Dallas office of Heidrick & Struggles and a key member of the firm's global Information Technology Officers Practice. Previously he was a principal at Spencer Stuart and an executive director at Russell Reynolds Associates. Steven is a member of the Association of Executive Search Consultants and the Society for Information Management). He earned his undergraduate degree in business from Jacksonville State University in Alabama, along with graduate studies at the University of Texas in Dallas.

Justin Lahullier, chief information officer, Delta Dental of New Jersey and Connecticut, joined Delta Dental of New Jersey in early 2000 and has held several key positions in areas including compliance, product development, strategic planning, and informatics. He was previously assistant vice president of Strategic Decision Support and Data Services, where his focus was on driving value using analytics, data visualization, and storytelling. He was promoted to CIO and VP of Information Services in April 2017.

Justin is a 25+ year veteran and past chief of the East Rutherford (NJ) Fire Department. He also serves as an appointee by the Bergen County Freeholders to a seat on the Public Safety and Communications Advisory Board. He holds a master's in health-care administration from Monmouth University and a postgraduate certificate from Drexel University in epidemiology and biostatistics.

Zackarie Lemelle is president/CEO of New Hope Coaching & Consulting. He is a visionary leader who has spent a lifetime using his passion for people, savvy for business, and an unshakable focus on excellence to transform leaders and organizations from the inside out. With over 40 years of experience working in firms, from startups, to Fortune 500s, to not-for-profit, Zackarie has honed his perspective and practice on what it takes to really nurture sustainable performance. He spent 18 years with Johnson & Johnson, where he served as vice president/CIO for Ortho-McNeil (now known as Janssen Pharmaceutical), Ethicon, and corporate headquarters. Then he successfully transferred those invaluable experiences to his pioneering role as a managing partner of one of the nation's top coach training firms. Now at the helm of New Hope Coaching and Consulting, his experience as a Certified Executive Leadership Coach (ACC–ICF), respected corporate leader, and purpose-driven business owner gives him a uniquely rich perspective on executive leadership and business transformation. Zackarie is also a current member and past chairman of the board for the Information Technology Senior Management Forum, a not-for-profit organization that has leaders from the top Fortune 50, 100, and 500 companies in the United States. He currently serves on the board of the Global Leadership Forum, a not-for-profit whose mission is to increase the number of disadvantaged children who choose STEM as a profession/vocation, a public-private partnership with the Congressional Black Caucus, Worldwide Technologies, ITSMF, BDPA, and Communication Careers Group.

Tony Leng is a managing director of Diversified Search and leads the firm's Digital Transformation and CIO practice from its San Francisco office. Previously he was managing partner of Hodge Partners. He has conducted many C-level searches for organizations where leaders are transforming/building organizations that are taking advantage of digital technologies. Prior to his executive search experience, he was a board member of three public companies and CEO of a $600 million public company. Before that, he ran a $1 billion division of a telephone company focused on corporate users and had responsibility for all data services and networking products. While at the telephone company, he was founder and chairman of its ISP and a board member of its two-million-subscriber cell phone subsidiary. He uses his operating experience from being a CEO, financial background as a CPA and Chartered Accountant, and technical knowledge from the work that he has done for tech companies to drill down and understand at a nuanced level what his clients are seeking to achieve as they build their teams. He has written numerous papers on technology leadership and the challenges facing technology leaders today and authors a blog, *Resources for the Modern CIO* (www.tonyleng.com). Tony has facilitated the Fisher CIO Forum for the past 12 years. The Forum, a monthly venue for CIOs and CTOs to discuss topics of relevance in their daily lives, affords these leaders the opportunity to develop a close network to discuss current challenges in a confidential peer group environment. He received a bachelor of Commerce (with honors), University of Cape Town, South Africa.

Beverly Lieberman is president of Halbrecht Lieberman Associates, Inc., an internationally recognized executive search firm founded in 1957. The firm provides retained executive search services across multiple industries while specializing in information technology. She joined Halbrecht Associates in 1986 and became the owner in 1991. She has successfully managed searches for Fortune 500 companies as well as early-stage businesses. Industries served include health care, insurance, higher education, financial services, managed service providers, retail, manufacturing, government, management consulting, high technology, law, and communications. Beverly is a frequent speaker at professional societies and conferences and is often quoted in major business and trade publications. She has been a featured writer for *CIO* magazine, is a long-standing member of the Society for Information Management, and currently serves on its board of directors. In addition, Beverly is a board member of the University of Bridgeport's Industry Advisory Board. She was selected by *Executive Recruiter News* as one of the 50 leading retained search professionals. She is also recognized as one of the nation's top recruiters by John Sibbald's The Career Makers. Beverly has a B.A. and M.A. from the University of Michigan and graduated from the prestigious MIT Sloan School Executive Management Program.

Ralph Loura is senior vice president and chief information officer at Lumentum, where he is responsible for digital strategy and operations for the company globally. Prior to joining Lumentum, he served as chief technology officer at Rodan + Fields. While there, he defined and advanced the company's technology strategy and infrastructure to support its significant growth and to deliver user-friendly digital experiences.

Prior to joining R+F, Ralph served as the CIO for HPE's Enterprise Group and HPE Labs, where he was a part of the core executive leadership team that led one of the largest corporate separations in history, splitting HPE from HP, Inc., ensuring continuity of service, compliance with separation agreements and regulatory requirements, and establishing a new cost and service base to support the two independent organizations going forward. He has received numerous industry awards, including *Computerworld*'s 2012 Premier 100 IT Leaders, Consumer Goods Technology's 2013 CIO of the Year, and HMG Strategy's 2017 Transformational CIO Leadership Award. Ralph is an angel investor, startup advisor, and founding board member of the TBM Council and has served as chair of the University of California Berkeley Fisher Center San Francisco CIO Forum. He holds an M.S. in computer science from Northwestern University and a B.S. in mathematics and computer science from Saint Joseph's College.

Christopher Mandelaris was formerly the vice president, chief information security officer, and chief privacy officer for Chemical Bank, responsible for information assurance, privacy, IT governance, and IT risk for the organization. He was most recently with Talmer Bank as managing director, chief information security officer. He started his career in IT Infrastructure Ford Motor Credit as a system analyst. He holds an

M.S. in information systems from Walsh College of Business and maintains several certifications, such as C | CISO, CISM, CRISC, CISA, PMP, MCSA, ITILv3, Six Sigma Greenbelt, MCP, and CNA.

Quintin McGrath is senior managing director, Technology Management & Enablement, Global Technology Services at Deloitte. He joined Deloitte over 20 years ago in South Africa as their first CIO. In late 2001 he moved with Deloitte to the United States as part of the first Global CIOs leadership team. Over the years, he has led various global technology programs, eventually leading the Global Application Development team for eight years. In 2015 he initiated the transition of Deloitte to becoming a shared services–based organization, and he now leads a team that is driving a more agile and responsive Deloitte Global IT. This includes driving global enterprise architecture, innovation, quality, and change. He is a long-standing member of Deloitte's Global Technology Executive team. Quintin holds an M.S. in engineering from the University of Cape Town and a business leadership MBA from the University of South Africa. As a proponent of lifelong learning, he is currently reading for an MA from Wheaton College in Chicago, focusing in part on effective cross-cultural leadership and communication.

Lynn McMahon is the managing director for Accenture in New York Metro. She is responsible for driving the local business strategy and engaging a staff of more than 4,700 New York– and New Jersey–based resources. In addition to her New York Metro responsibilities, she leads the Media and Entertainment industry group for Accenture in North America, a practice that serves clients in the broadcasting, entertainment, internet, social, and publishing industries. In 2000 she founded the Accenture Innovation Center in New York Metro and subsequently served as its director for six years. This center, the first of its kind when founded, provides leading-edge technology capabilities and solutions, demonstration and development facilities, skilled professionals, and cross-industry expertise to help companies bring their ideas to life quickly. Lynn is the founder and executive sponsor of the Accenture Women's Leadership Forum, a premier client event for senior female business executives in the communications, media, and high-tech industries. She is a recipient of Accenture's prestigious "Great Place to Work for Women" award, which is given annually to a senior Accenture executive who has fostered and modeled an environment where women can succeed. She holds a bachelor's degree in finance and a master's in business administration, both from Florida State University. She serves as a vice president on the board of directors of the New Jersey Ballet Company and on the board of the World War II Memorial in Washington, DC.

Matt Mehlbrech is vice president of Information Technology at CoorsTek, Inc., a global leader in engineered ceramics, serving numerous industries, from semiconductor manufacturing, to energy and defense solutions, to medical devices. In his role, he leads the definition and execution of the information technology strategy for CoorsTek, partners with the various business units, manages all IT resources, and supports all business systems and processes, including ERP and manufacturing-related systems, global infrastructure, information security, and end user tools. Prior to joining CoorsTek, he spent over 18 years with Eaton Corporation, a $20 billion power management company providing products and services in the electrical, vehicle, hydraulics, and aerospace industries. Matt holds a B.A. in business administration and Honors Program in management information systems from the University of Wisconsin–Milwaukee.

Vipul Nagrath is global chief information officer for ADP, with responsibility for all of ADP's infrastructure, operations, enterprise applications, enterprise technology systems, and processes across the globe. Vipul and his team work closely with the company's senior business leaders to execute ADP's business strategy and identify, prioritize, and execute technology solutions that are cost effective, efficient, and scalable globally. The focus of Vipul's organization is working collaboratively across the organization to ensure that strategic capabilities that support an excellent service experience for clients and associates are built. An industry leader in technology, Vipul brings more than 25 years of experience in developing forward-thinking solutions. Prior to joining ADP, Vipul was with Bloomberg L.P., where he started as a project manager in 2003 and rose to become Bloomberg, L.P.'s global head of Research & Development. He continued his growth and responsibility as CIO of Bloomberg's $900 million Enterprise Solutions Group business, where he had full oversight of all operational and technology needs for the enterprise business globally. In his prior role as the global head of R&D, Vipul oversaw technology research and development for fixed income, foreign exchange, equity, trading solutions, commodities, options, derivatives, data, portfolio analytics, internal systems, web products, software infrastructure, and news.

Vish M. Narendra is senior vice president and chief information officer of Graphic Packaging Holding Company, a position he has held since May 2015. He joined the company from General Electric Company, where he worked from September 2001 to April 2015, serving most recently as CIO for the Alstom integration for the GE Power & Water business. From September 2001 to August 2014, Vish held a series of roles with increasing responsibility, including a stint as the Asia region CIO for the energy and aviation businesses within General Electric Company, CIO for demand management within the power generation business, and other functional roles. From 1999 to 2001, Vish was with Idea Integration, a provider of digital strategy and application development. Vish holds an undergraduate degree from Anna University, College of Engineering, Chennai, India, in electrical and electronics engineering and an MBA from the Stuart School of Business, Illinois Institute of Technology, Chicago.

Earl Newsome is chief information officer of the Americas IT organization for Linde, a global industrial gases company with more than 80,000 employees in more than 100 countries and a 2018 revenue of $28 billion. In his current role, he provides leadership and strategic vision for an aligned Americas IT organization within the corporate IT strategy framework, drives key IT projects, and supports the company's standardization initiative, with a focus on harmonizing and replicating wherever possible. Another key priority is cultural integration so that diverse teams work together more effectively. Prior to joining Linde, Earl served as corporate CIO and vice president, Digital, at technology company TE Connectivity. In this role, he built a world-class international IT organization that accelerated the success of each business-to-business unit. He serves on the boards of directors of First Independence Corporation and First Independence Bank, Detroit. Earl holds a bachelor's degree in computer science from the United States Military Academy in West Point, New York.

Helmut Oehring is the executive vice president of Asteelflash. He is an executive thought leader with expertise in global operational efficiencies, design, and implementation of strategic IT initiatives and projects producing stellar bottom-line results. He has an extensive background in ceramics, oil and gas, health care, and defense industries. A visionary and implementer of process improvements using best practices to build high-quality organizational performance through dynamic leadership, Helmut also is a strategic and innovative thinker with success driving profitability through leveraging technology and business intelligence to transform operations.

Tom Peck serves as executive vice president, chief information and digital officer for Ingram Micro Inc., where he oversees all aspects of the company's Global Information and Digital Technology efforts. In this dual role, he guides the strategic development and evolution of Ingram Micro's technology backbone, leading a growing team of engineers and technical staff who support the technology, systems, and digital requirements of the company's global business operations. His immediate areas of focus include developing a long-term systems architecture and effectiveness plan designed to deliver additional speed, agility, and scalability to support both internal and external partners. Tom's global perspective, leadership, and vision will help the company continue to deliver an enhanced customer and associate experience by modernizing and digitizing core systems and processes and integrating them across all internal and external-facing products and platforms. Prior to joining Ingram Micro in March 2018, he most recently served as senior vice president and global chief information officer at AECOM, a multi-billion-dollar global provider of professional technical and management support services. In this capacity, he was responsible for the support and deployment of technology used to serve more than 100,000 employees and thousands of clients across 150 countries. He has also occupied high-performing CIO leadership roles at Levi Strauss & Company, MGM Resorts (formally MGM MIRAGE), and General Electric–owned NBC Universal's global entertainment business unit. During his tenure at NBC Universal, he also served as quality leader and was responsible for process improvement and global Six Sigma deployment. Tom began his career as an officer in the United States Marine Corps, holding numerous finance and technology roles inclusive of both operational and headquarter assignments and culminating in an enterprise-wide program management role on the Marine Corps chief financial officer's team at the Pentagon. He holds an M.S. in management from the Naval Postgraduate School, a B.S. in economics from the United States Naval Academy, and is a certified Six Sigma Master

Black Belt. In 2015 he was inducted into the CIO Hall of Fame and currently sits on the board of directors of Veterans Park Conservancy as CFO and treasurer.

Wendy Pfeiffer is chief information officer of Nutanix, where she focuses on enterprise adoption of modern technologies to fuel the company's global mission. She also serves on the boards of Qualys, Inc., and Girls In Tech. A consumer tech enthusiast, Wendy has led technology and operational functions for Robert Half, GoPro, Yahoo!, and Cisco. Her recent accolades include being named ORBiE's Bay Area Enterprise CIO of the Year, the Fisher Center for Data Analytics' CIO of the Year, and one of HMG's Top Technology Executives. She was also ranked first on Enterprise Management 360's list of Top 10 Tech CIO's, was named one of *Silicon Valley Business Journal*'s Women of Influence, and was listed as one of the National Diversity Council's Top 50 Most Powerful Women in Technology.

Steve Phillips is the chief information officer at Alorica, where he is responsible for setting and delivering IT strategy. Between 2005 and 2017 he served as CIO at Avnet. From 2004 to 2005 he was CIO at Memec, a global electronic components distributor acquired by Avnet. Since 2012 he has served as chairman of the board of Wick Communications, a U.S.-based news media business. In 2015 Steve was inducted into the CIO Hall of Fame, a lifetime achievement that honors IT executives and visionaries whose leadership has significantly impacted the field of IT. He holds a bachelor's degree in electronic engineering from Essex University and a postgraduate diploma in management studies from the University of

West London. He is a fellow of the Institution of Engineering and Technology.

Steve Phillpott was named chief information officer for Western Digital Corporation in October 2015. Prior to this, he was CIO for HGST, a Western Digital brand. As the CIO of Western Digital, he oversees the worldwide IT department encompassing infrastructure and operations, global applications, enterprise data management and IT security. Steve works with leaders to provide strategic and tactical IT direction, which often includes utilizing Cloud, SaaS, Social, Web, Mobile and Big Data to drive business innovation and value. Based in the U.S. headquarters in Irvine, Calif., he manages WDC's global, multi-disciplined IT team.

Prior to joining HGST, Steve was CIO for Amylin Pharmaceuticals, a leading provider of drugs for the treatment of diabetes. Steve was previously the CIO of Proflowers.com, a high-volume e-commerce retailer, and the VP of IT for Global Enterprise Applications at Invitrogen (now Thermo Fisher Scientific). Prior to that, he held various leadership positions at companies including Memec (Avnet), Gateway and Qualcomm.

A recognized industry expert, Steve has received numerous industry awards. Most recently, he was recognized among Computerworld's top 100 IT leaders for 2016. Additionally, he was named a finalist in the Public Company category in the Bay Area CIO of the Year Awards for 2016. He is a frequent speaker on IT innovation, transformation and cloud migration strategies at many leading industry events, including Amazon Executive Summit, and events sponsored by Gartner, Forrester, BIO-IT World, Information Week 500 and numerous others.

Steve received his BS in Engineering from the U.S. Naval Academy in Annapolis, Maryland, and holds an MBA in Technology Management from the University of Phoenix.

Mark Polansky is a senior partner in Korn Ferry's Technology Officers Executive Search Practice, which he cofounded and headed for more than 12 years. He extensively recruits chief information officers, chief technology officers, chief digital and chief data officers, chief information security officers, and other senior IT leadership talent across vertical sectors including industrial, consumer, life sciences, technology, and higher education. Before entering the search field, Mark developed and marketed information systems in the financial services and higher education sectors. He currently serves on the Advisory Committee for Columbia University's executive graduate program in Information Technology Management. He frequently addresses conferences and writes on information technology and career management topics. He originated the "Executive Career Counsel" column in *CIO magazine* and on CIO.com. Mark is a member of the Society for Information Management and served as president of the New York chapter. He has served on the advisory boards of The Information Technology Senior Management Forum, the national organization dedicated to fostering executive talent among African American IT professionals, and HITEC, the Hispanic Information Technology Executive Council. He holds a master's degree in computer science from Pratt Institute and a bachelor's degree in mathematics and electrical engineering from Union College.

Phyllis Post is the former vice president and CIO for Merck's Global Human Health division, where she was responsible for enabling business value through a strategic partnership of business and IT aligned to realization of critical priorities of the corporation. This included design, delivery, and implementation of both foundational and innovative, leading-edge technology products and platforms to support enhanced customer engagement, value-add automation, and robust analytical insight and decision support. In her previous role as the Asia Pacific Client Services and IT Regional Hub leader, Phyllis provided leadership to establish MSD's IT Regional Innovation & Development Center in Singapore while also serving as the regional leader for the IT business partnerships for the commercial, manufacturing, and R&D divisions. Prior to joining Merck, she worked for a variety of mid-tier typesetting and web development companies that were engaged primarily in the publication of financial, scientific, medical, and technical materials. Additionally, she served as a board member for the Typesetters Association of NY and as president for PrintNJ. She holds an MBA in technology management from the University of Phoenix and a B.A. from Douglass College.

Nathalie Rachline is a senior executive in IT; she has led major transformations and negotiated and managed large outsourcing deals implementing strategic and business critical partnerships, most recently at Nutrien (ex. Agrium). She is also a specialist in technology business management and has managed IT operations and strategic planning and governance in many vertical domains in France and the United States. She has also run an IT transformation and service management consulting practice inside a major managed service provider and has successfully delivered on several mission-critical strategic consulting engagements. Nathalie is on the board of the Colorado Chapter of the Society for Information Management and is also very active in the entrepreneurial and startup community in Northern Colorado as mentor and advisor.

Dan Roberts is president and chief executive officer of Ouellette & Associates. He has authored and is a contributing author of numerous books. His latest, *Unleashing the Power of IT* and *Confessions of a Successful CIO*, were on Amazon's list of top-rated books for months and are being leveraged by IT leadership teams as models for moving IT up the maturity curve. A keynote speaker and panel moderator at frequent industry conferences and corporate events, he is considered one of the best-connected thought leaders in the CIO space. Dan has a passion for bringing CIOs together to learn and benefit from one another's experiences. Each year he works and meets with more than 1,000 CIOs and IT leaders across the globe. As the CEO and president of Ouellette & Associates Consulting, Inc., he leads the firm known since 1984 for Developing the Human Side of Technology. His team, along with a cadre of ecosystem partners, has helped more than 3,500 IT organizations build world-class cultures, high-performing workforces, and differentiated talent brands. In 2017 he founded HR2IT, an exclusive community for HR executives who support the unique needs of the CIO and IT leadership team. HR2IT is committed to helping its members build an IT workforce strategy that differentiates their organizations in the marketplace and equips them to execute their company's digital transformation journey. Dan has been an active member of the Society for Information Management for more than two decades. In 2018 he was recognized with the distinguished "Leader of the Year" award, having been instrumental in launching the new Tampa Bay chapter.

Dave Roberts is CIO at Radius Payment Solutions, responsible for the in-house IT development and support teams at the group's headquarters in Crewe and the Manchester-based Technology Centre. Under his leadership, Radius Payment Solutions is excelling on delivering advanced industry-leading technology that differentiates the organization in the fuel card and telematics markets. Dave has worked in the IT industry since 1997, with previous director experience across global IT teams. He is committed to delivering transformational IT services and driving forward programs of change. He participates in a number of IT advisory boards, providing recommendations to groups within education, industry, and government. He was appointed as a fellow of BCS, The Chartered Institute for IT, in 2015. He has a B.S. degree in business information systems and an MBA.

Jon Roller is chief information officer at Horsley Bridge Partners, a private equity partnership (Fund of Funds). He combines his accounting, finance, and information system experience managing all IT and IS functions across the firm's global offices and developing decision-making and investment management resources. Beginning in the early 2000s, he worked to develop the firm's investment data systems and since 2009 has controlled all design and development of those systems. Jon's responsibilities include oversight of teams of accountants and international software engineers, tasked with data capture. He graduated from Miami University in Oxford, OH, with a B.S. in business finance with an emphasis in political science.

John Rossman is managing partner, Rossman Partners. With more than 28 years of technology strategy, design, implementation, and operating experience, he has led several complex businesses and programs resulting in innovative business models. He has worked with clients across a broad range of industries, including retail, insurance, education, forest products, industrial products, and transportation. He is the author of *The Amazon Way: 14 Leadership Principles of the World's Most Disruptive Company* and is an expert on digital disruption and assisting his clients to build and execute new business models.

John was director of Enterprise Services at Amazon.com, managing worldwide services to enterprise clients such as Target.com, Toys R Us, Sears.ca, Marks & Spencer, and the National Basketball Association. In this role, he had full operational and technical ownership for existing clients, overseeing e-commerce solutions, such as online merchandising capability, website technologies, branded fulfillment delivery, and branded customer service. He earned a bachelor's degree in industrial engineering from Oregon State University.

Kenneth Russell is the former vice president, Digital Transformation, and chief information officer at Pfeiffer University. He is a respected organization and technology leader/coach and turnaround specialist. He is an organization builder (North Carolina, Silicon Valley, Kansas) and has more than 18 years of experience developing and implementing comprehensive organization change and transformation efforts, bespoke programs, and innovation initiatives. Kenneth is known for successfully leading and stabilizing turbulent environments. A technology pioneer, he developed early intranet systems for large banks, which led to his participation in developing one of the first successful internet-based training platforms. He recently completed a comprehensive digital transformation and organization change effort at Pfeiffer University, serving as the University's CIO and vice president of Digital Transformation. Kenneth is a founder and president emeritus of both the Charlotte and Wichita chapters for the Society for Information Management. He was a director and strategist at Cisco, where he led the Intellectual Capital Transformation group.

Vijay Sankaran is chief information officer of TD Ameritrade, where he is responsible for incubating, introducing, and managing new capabilities as well as for the implementation of the company's data and analytics ecosystem, including bringing enhanced personalization to the client experience. Vijay also serves as a member of the company's senior operating committee, which shapes the strategic focus of the organization.

Prior to joining TD Ameritrade, he held several senior leadership roles at Ford Motor company, including: IT chief technology officer, leading advanced technology, technology strategy, data and analytics, and architecture across the enterprise; director of application development, with responsibility for global application development and the delivery of major programs and projects; and director of infrastructure operations, with responsibility for operational support for infrastructure as well as Ford's largest data centers in North America. He holds a bachelor's degree in computer science from the Massachusetts Institute of Technology and an MBA from Duke University's Fuqua School of Business.

Candida Seasock, founder and president of CTS Associates, LLC, is an innovative business advisor focusing on mid-size companies specializing in enabling client growth and management success through internal executive teams and/or advisory boards that support vision, innovation, change that supports operation, finance, and technology and employee growth. As a professional mentor, she focuses on businesses that are interested in strategic growth. Specializing in enabling client growth and management success, she has developed her own award-winning approach, "Growth Path to Success" based on targeted business development, strategic marketing, fostering strong and long-term client relationships, and implementing 360-degree agility into a company's processes and operations. She has successfully assisted management teams ranging from Fortune 500 corporations to emerging growth companies. Candida is best known for her work in building high-value market recognition for mid-size companies.

Bhavin Shah is the founder and CEO of Moveworks, a cloud-based AI platform, purpose-built for large enterprises, that solves one big frustrating problem: resolving employees' IT support issues. Bhavin is a three-time entrepreneur with over 18 years of experience in building companies across different industries—consumer, mobile productivity, and enterprise AI. He combines a technical and creative mind-set, leading teams from ideation to product delivery and from inception to outcome. Prior to Moveworks, he was the CEO and founder of Refresh.io, which was acquired by LinkedIn. He started his career in the toy and video game industry. Moveworks uses a strong form of AI to resolve enterprise IT issues autonomously, with Fortune 500 and large enterprise customers, such as Autodesk, Broadcom, Medallia, Nutanix, and Western Digital.

Naresh Shanker is senior vice president, chief technology officer for Xerox Corporation. In this role, he is leading Xerox through its digital transformation, responsible for research and product development including supporting the company's product portfolio with PARC from its ideation to commercialization phases and an increasingly digital information technology operation. He was appointed to this position effective May 6, 2019, and a senior vice president of the corporation on May 21, 2019. Naresh's experience within the IT industry spans more than 25 years. Most recently he was chief digital and information officer for a startup company focusing on disruptive nano-materials and clean energy solutions; he continues to be a strategic advisor for that firm. Previously he was the CIO for Hewlett Packard and Palm, Inc. He serves on the advisory boards of several companies, including Devo, Inc., Cylance, Inc., and AutonomIQ, Inc., and is on the board of directors for Clarizen. He earned his MBA and B.S. in computer science from the Illinois Institute of Technology in Chicago.

Prabhash Shrestha is executive vice president and chief digital strategy officer for the Independent Community Bankers of America®. In his role he leads the association's information technology and digital efforts, ensuring their alignment with ICBA's business goals and objectives. He has more than 20 years of information technology experience, including leadership roles with the Association of Fundraising Professionals and the Association of Trial Lawyers of America. He most recently served as vice president of technology for the American Gastroenterological Association, where he led the organization's technology strategy and operations. Prabhash has received a number of industry honors, including a Top Association and Non-Profit Innovators award in 2017 and 2015, a 2016 Top Association Tech Guru award, and several teaching excellence awards at Georgetown University. He is a member of the governing board of the Wise Giving Alliance and chair of its technology committee. He is also a technology committee member for the Council of Better Business Bureaus. Prabhash holds an MIT Sloan Executive Certificate on Strategy and Innovation and has entrepreneurship certificates from the Wharton School of Business. He is a professor at Georgetown University, where he teaches a master's program on technology management. He has a B.S. and an M.S. in information systems from Strayer University and is an ASAE Certified Association Executive.

Patrick Steele, chair, CIO Advisory Board, Blumberg Capital, is an innovative executive with experience in all aspects of health care, insurance, retail, wholesale, supply chain, merchandising, information technology, dot.com profit and loss, and worldwide retail exchange functions. Patrick is a technology-astute executive with generalist leadership talent, focused on the intersection of technology and business, from a strategic perspective. He is a proven executive with the ability to create a vision, secure the monetary resources, and focus the management team to align the human capital needed to achieve the strategic execution that brings a vision to life. As a member of executive management teams, he successfully led several business process reengineering and technology refresh projects. He is an excellent communicator and speaker in presentations to both boards of directors and the public. Patrick holds a B.A. in business administration and a B.S. in mathematics, both from the University in Washington.

Milos Topic is vice president and chief information officer at Saint Peter's University. He has 20 years of experience in leadership, innovation strategies, technology implementation, and business development. His formal education is a blend of science, technology, and business. Milos's journey in the IT profession started in 1997, and over the past 20+ years he has worked on nearly all aspects of IT, including networking; cabling installs; tech support; programming in C++, C#, and Java; web development; and system/network security/administration. His most recent positions involve leading teams of amazing people providing technology solutions and services while supporting a multitude of organizational needs.

Angie Tuglus is a senior advisor and technical fellow at AQN Strategies, CGS Advisors LLC. She formerly served as executive vice president and chief operating officer at Ally Insurance, where she oversaw all insurance operations and led their new direct-to-consumer business. Prior to Ally, she began her career in technology startups and then worked at Ford Motor Company in Global Technology and Consumer Insights.

Sangy Vatsa is executive vice president and chief information officer of Comerica Bank, responsible for all digital technology, cybersecurity, enterprise data office, business continuity, and technology risk management at the bank. Prior to joining Comerica Bank, he served as the CIO for global customer service, credit, collections, and payments issuance at The American Express Company. At American Express, he served in leadership roles for IT strategy, architecture and innovation, enterprise shared services, global business travel, and corporate payments digital capabilities. Sangy holds an MBA from Ross School of Business, University of Michigan, a master of computer and information systems degree, and a bachelor of electronics and telecommunications engineering degree.

Ken Venner is the former CIO of SpaceX, where he was responsible for overseeing the design, development, and implementation of its state-of-the-art computing and information sharing infrastructure. He went to SpaceX from Broadcom, where he served as chief information officer. Prior to Broadcom, Ken was vice president of Product Management and chief information officer of Rockwell Electronic Commerce. He received a bachelor of engineering degree from the Stevens Institute of Technology, a master of engineering degree from Carnegie Mellon University, and an MBA from New Hampshire College.

Gautam Vyas is executive vice president and division executive at FIS Global. He was recruited to FIS in 2016 to build its Global Professional Services business and accelerate its growth trajectory. He added growth officer responsibilities in 2018 and expanded his remit to build and grow services business for the Banking Solutions segment in 2019. He now leads the Banking Solutions Services business spanning digital and banking, payments and wealth. As global head of Professional Services, Gautam's responsibilities include accelerating FIS's top- and bottom-line growth through new and unique services, productizing services, go-to-market through expansion of client footprints, and creating new distribution channels. The Professional Services division of FIS is a digital and banking and payments modernizer that actively supports financial institutions globally, helping clients meet their unification and transformation goals across the entire organization, through strategy consulting, business transformations, proactive and bespoke services, system integrations, implementation of software and ancillary services, and go-to-market plans. Gautam earned his M.S. in electronics from S.P. University and his MBA from Kansas State University.

Craig Walker is vice president and chief information officer of Shell Downstream, Shell International Petroleum Company. He joined Shell Downstream as global CIO in July 2014. He came to this role following an assignment in Shell Trading and Supply, where he served for six years as the global CIO based in Houston, TX. He is also a member of the RDS IT Executives. He has nearly 28 years of experience with Shell, in both the upstream and the downstream IT functions. During this time, he has had overseas assignments in Colombia, Dubai, Saudi Arabia, South Africa, the United States, and the United Kingdom. Prior to joining Shell, he spent five years with KPMG Consulting, where he managed strategy, large system implementations, and process change programs for major global clients.

Bart Waress is vice president of information technology at Discovery Natural Resources. He is a peer-awarded (2018 IT Leader of the Year) strategic IT leader focused on business technology uplift and automation and has been recognized for his ability to align IT's roadmap with the C-suite vision. He has led IT organizations in various industries, including oil and gas, health care, banking, finance, and telecom. Bart has a proven background in IT leadership that includes both managing and creating corporate automation, cybersecurity organization, data and system architecture, technology selection and architecture, cloud transitions, IOT, project management, and budget management. He has a strong understanding of financial impacts that technology and IT can have for a business.

Patricia Watters is managing partner in DHR International's Detroit office and global leader of DHR's Automotive and Future Mobility Practice Group. She brings over 20 years of unique and diverse leadership in automotive manufacturing and manufacturing consulting to the executive search community. Prior to joining DHR, she spent six years as a vice president of Harbour Consulting, where she led engagements in research, manufacturing strategy, and operations improvement for automotive and nonautomotive clients and was a principal contributor to the leading industry publications, such as *The Harbour Reports*, which analyze and rank assembly, stamping, and powertrain plants in productivity and lean manufacturing performance. Patricia holds a B.A. degree from Kalamazoo College with a major in economics and a minor in international business and an M.A. degree in industrial relations from Wayne State University.

Jon Wrennall is group chief technology officer at Advanced, where he is driving innovation, leading the Product and Engineering teams to deliver products and services that make the complex simple and make a difference for Advanced's customers and the broader society. It is his responsibility to bring innovation and quality in the development of Advanced's suite of software solutions and services to make a difference to its 20,000 customers. A renowned thought leader in the technology sector, Jon has over 25 years of experience in the industry. Prior to Advanced, he was CTO for Fujitsu UK&I for five years, where he helped transform the 1,200-strong architecture community. As a Fujitsu fellow, he re-created the Fujitsu Distinguished Engineer scheme. He was a founding and core member of the UK Government CTO

Council and recruited and led a team creating Public Services Network, XBRL mandating and cross-government channel strategy. He is a Chartered Engineer, fellow of the British Computer Society, and sits on the Innovation Board of the CBI.

Angela Yochem is executive vice president and chief digital officer for Novant Health, a superregional health care system with one of the largest medical groups in the United States. She and her teams deliver the world-class consumer capabilities, differentiating technologies, and advanced clinical solutions that allow the high-growth system to provide remarkable patient care. Angela has served as EVP/CIO at Rent-A-Center, global CIO at BDP International, global CTO at AstraZeneca, and divisional CIO at Dell. She has held tech exec roles at Bank of America and SunTrust and held senior technology roles at UPS and IBM. In these roles, she built business-to-business digital product lines, grew digital retail channels (B2C), created technical services lines of business, and transformed global technology capabilities. Angela has been a director for the Federal Home Loan Bank of Pittsburgh, BDP Transport, BDP Global Services Asia and Europe, and Rocana, with experience on audit, enterprise risk, operational risk, and governance/policy committees. She remains an EIR for Vonzos Partners, a mentor for SKTA Innopartners, and an advisor for Dioko Ventures. She serves on the board of Freedom School Partners, a nonprofit committed to promoting literacy in the Charlotte area, and on the executive team of the Go Red for Women organization, part of the American Heart Association. She is a trustee of the Charlotte Regional Business Initiative and is an advisory board member for the American Hospital Association Innovation Council and the

University of Tennessee Electrical Engineering and Computer Science department. Angela has a bachelor of music degree from DePauw University and an M.S. in computer science from the University of Tennessee. She holds three U.S. patents and is an author with Addison-Wesley and Prentice-Hall.

Eric Yuan is founder and chief executive officer of Zoom. Prior to founding Zoom, he was corporate vice president of Engineering at Cisco, where he was responsible for Cisco's collaboration software development. As one of the founding engineers and vice president of Engineering at WebEx, he was the heart and soul of the WebEx product from 1997 to 2011. Eric is a named inventor on 11 issued and 20 pending patents in real-time collaboration. In 2017 he joined the Forbes Tech Council and was added to the *Business Insider* list of the 52 Most Powerful People in Enterprise Tech.

Jedidiah Yueh is founder and chief executive officer of Delphix. He has led two waves of disruption in data management, first as founding CEO of Avamar (sold to EMC in 2006 for $165 million), which pioneered data de-duplication and shipped one of the leading products in data backup and recovery, with over 20,000 customers and $4 billion in cumulative sales. After Avamar, he founded Delphix, which pioneered data virtualization, earning over $200 million in its first seven years of sales. In 2013 the *San Francisco Business Times* named Jedidiah CEO of the Year. Well connected in the venture community, he has raised over $150 million from over 10 VCs and has more than 25 patents in data management. After being designated a U.S. Presidential Scholar by George H. Bush, he graduated Phi Beta Kappa, magna cum laude, from Harvard.

ABOUT THE AUTHOR

Hunter Muller is president and chief executive officer of HMG Strategy, LLC, a global IT strategy consulting firm based in Westport, CT. He leads the world's largest network of IT leaders and executives, providing foundational guidance for driving revenue growth and creating value for customers.

HMG Strategy operates a unique social-mobile-digital platform that enhances and accelerates the exchange of valuable timely knowledge across a global network of technology executives, vendors, developers, investors, and thought leaders.

Hunter has more than three decades of experience in business strategy consulting. His primary focus is IT organization development, leadership, and business alignment. His concepts and programs have been used successfully by premier corporations worldwide to lead, reimagine, and reinvent strategy and to drive alignment across the topmost levels of management.

ABOUT HMG STRATEGY

At HMG Strategy, we are committed to helping technology executives to be transformational leaders.

HMG Strategy is the world's largest independent and most trusted provider of executive networking events and thought leadership to support the 360-degree needs of technology leaders. Our regional CIO and CISO Executive Leadership Series, newsletters, authored books, and digital Resource Center deliver proprietary research on leadership, innovation, transformation, and career ascent.

The HMG Strategy global network consists of over 400,000 senior IT executives, industry experts, and world-class thought leaders. Additionally, our partnerships with the world's leading search firms provide vital insights into the evolving roles of the CIO and CISO.

INDEX

Garbage in, garbage out
(GIGO), 146
General data protection regulation
(GDPR)
absence, 125
usage/compliance, 43–44
Generous, Curtis, 167
Global economy
factors, 105–107
technology, relationship, 107
Global energy company, creation,
112–113
Global IT team, strengths, 73
Global shifts, confrontation, 123
Go-to-market activities, 61
Go-to-market vision, description,
84–85
Gray, Chuck, 16
Greenlots, business
(growth), 116
Groves, Leslie, 107

Hackathons, sponsorship, 210
Harding, Jon, 72–73
Hartnett, Kevin, 188
Hassan, Vishwa, 10–11
Hatter, Patty, 56, 57
Hawking, Stephen, 130
Hengels, Jason, 125–126
Herlihy, Donagh (interview),
160–165
HMG Strategy events, 16, 19, 26,
54, 126, 212
Holmes, Ryan, 146
Horizontal thinking process,
128
Human resources (HR), evolution
(theme), 8
Humility, display, 67

Ideas, communication, 130
Image recognition, 46
Immelt, Jeff, 195

Industrial Revolution, electric
dynamo (impact), 48
Industries
digital transformation, 170
events/conferences, importance,
20–21
technology, impact, 13
transformation, 39–40
Information technology (IT)
AIOps, usage, 158
business enabler role, 98
career paths, 174
consumerization, 21–22
executive leadership, excellence
(achievement), 9
IT-related functions/processes,
autonomics (design), 155
sending level (2019-2022), 8
shift, 9
support problems, solutions,
49–50
team
engagement, quality, 34
members, empowerment, 209
transformation, artificial
intelligence (impact),
130–132
understanding, 56
value driver, belief (validation),
136–138
value proposition, renewal, 74
Information technology service
management (ITSM), 157
Infrastructure, shift, 9
In-house awards program,
development, 210
Innovation
business survival, comparison,
191
competition, impact, 195
diversity, relationship, 209–210
Facebook monopoly, impact,
142–144